Good Fire Journaling

Through Pyrosketchology Practices

Written and Illustrated by
Miriam H Morrill

This is a privately authored and illustrated book developed for prescribed fire education and nature journaling purposes. The book synthesizes and stylizes a range of federal wildland fire training and educational materials for the purpose of observing and journaling prescribed and cultural burns.

Cover art and book illustrations by Miriam Morrill

Published by Morrill Intentions Press

979-8-9901071-2-0

Table of Contents

Introduction & Approach

To truly observe fire is to enter into a relationship with place. Fire behavior is not simply a function of fuel, weather, and topography; it is also shaped by perception, memory, and lived experience. Through the deliberate act of journaling, we slow our attention and begin to notice the subtle exchanges of heat, moisture, wind, and light that shape the fire environment.

This book is a supplement to *Pyrosketchology: An Illustrated Guide to Nature Journaling About the Fire Environment* and focuses on observing and journaling prescribed and cultural burning events. The goal is not artistic perfection but perceptual clarity. By combining field-based sketching with structured observation, readers can develop a deeper understanding of how fire moves through landscapes, how environmental conditions influence fire behavior, and how fire shapes our connections to place.

Many of the approaches presented here draw from established fire-management practices and training materials while incorporating techniques from nature journaling and field ecology. The intention is to create a bridge between operational knowledge and experiential learning. When observation becomes habitual, situational awareness strengthens, and both safety and understanding improve.

Although technology provides increasingly sophisticated tools for monitoring fire environments, direct observation remains irreplaceable. Instruments can measure temperature, humidity, and wind speed, but they cannot fully convey the sensory context in which fire operates. The warmth of a slope at mid-afternoon, the dryness of grasses underfoot, or the shifting texture of wind through tree canopies all provide meaningful clues that support interpretation of fire conditions.

This text is designed as a hybrid field guide—part instructional manual, part observational companion. It combines peer-reviewed fire science, agency training materials, and field-based observation practices to support a holistic understanding of good fire. Intended for fire practitioners, land managers, students, and nature observers, the book invites readers to learn through direct engagement with fire and the environments in which it occurs.

How to Use This Guide

This guide is intended to support both structured study and flexible field practice. You may choose to follow the chapters sequentially or move between sections based on seasonal conditions, training needs, or personal interests. The book begins with journaling approaches for Good Fire observations and an introduction to the burn frameworks, goals, objectives, and indicators that help guide what we observe.

The chapters that follow are generally organized around the three primary components of the fire environment—landform and place, fuels and vegetation, and weather. However, these elements rarely operate in isolation. In the field, terrain influences weather, weather influences fuels, and fuels interact with both. For this reason, some observations, journaling techniques, and exercises intentionally blend multiple elements of the fire environment. They are introduced where they are most strongly experienced, observed, or understood. A story map of mountain landscapes and cloud development, for example, may appear in the Landform & Place chapter because it helps reveal how terrain shapes the experience of place, even though it also relates to weather. Throughout the book, concepts are revisited and expanded as new layers of observation and understanding are added.

Each chapter combines foundational fire-science concepts with observational exercises designed to strengthen situational awareness, support fire-effects interpretation, and deepen relationships with the landscapes where Good Fire occurs.

Some readers may approach this text from an operational fire-management perspective, while others may be motivated by ecological curiosity or creative journaling practices. Regardless of your starting point, the exercises are designed to be adaptable. Observations can be made during prescribed burns, training simulations, or independent field exploration.

The journaling approaches presented here emphasize repeated practice over technical precision. Sketches may be simple, symbolic, or highly detailed depending on time constraints and observational goals. Written notes may include measurements, sensory impressions, questions, or reflections. Over time, these records become a personal archive of environmental patterns and fire behavior insights.

Observational Ethics and Safety

Observing good fire requires careful attention to safety and professional boundaries. This guide does not replace formal training, operational protocols, or incident command structures. When participating in prescribed fire, always follow agency guidelines and direction from qualified leadership.

Independent observers should maintain safe distances from active fire and avoid interfering with management activities. Ethical observation also includes respect for land ownership, cultural sites, and ecological sensitivity. Journaling practices should enhance situational awareness rather than create distraction or risk.

In many cases, meaningful observations can be made outside active burn perimeters by focusing on weather patterns, fuels, terrain influences, and post-burn fire effects. These indirect observations contribute significantly to understanding the knowledge, skill, art, and science behind prescribed fire.

Developing Observation Skills

Developing observation skills depends less on artistic ability or prior fire science knowledge and more on consistent attention to targeted observations. Recording notes, sketches, and field measurements helps strengthen pattern recognition and interpretive skills over time.

With repeated practice, observers become better at recognizing environmental patterns and anticipating changes in weather, fuel moisture, fire behavior, and fire effects. Establishing regular observation routines—such as morning weather checks, seasonal landscape mapping, or post-fire site visits—helps connect field observations with environmental change and fire behavior.

Foundational Journaling

I recommend a foundational technique to describing, organizing, and processing observations that was developed by John Muir Laws . This includes the use of Words, Numbers, and Pictures to describe observations in a journal. Each of these "languages" uses a different part of your brain to process information, creating a more full-brained learning approach.

- **Words**—narrative descriptions of what you notice
- **Numbers**—quantitive measurements to describe what you notice and or see patterns or trends
- **Pictures**—visuals that represent and accentuate what you notice, such as sketches, diagrams, maps, tables, and graphs. Although some photos can be used, creating the visuals is the better process for clarifying what you notice. This is also a process of visual thinking and by using symbols and color codes with numbers you can add deeper relevance.

Good Fire Journaling Approaches

There is no single way to journal about fire. The observations you make, the details you record, and the way you organize your journal often depend on your purpose and time available. Sometimes you may want to explore the relationships between people, place, and fire. Other times you may want to better understand fire behavior, smoke movement, or ecological effects. In operational settings, you may be focused on recognizing changes that affect safety and decision making.

This book uses three complementary journaling approaches to help guide observations and organize information: the Story Approach, the Study Approach, and the Situational Awareness Approach. While each has a different emphasis, they often overlap and can be used together during the same fire event.

Observation Scales

Another important consideration for journaling about fire is the relevant scales of observation. There are temporal and spatial scales that are more relevant for certain types of observation and data. For example, a larger landscape sketch or map is more relevant if you want to observe the relationship between wind and landform, whereas fire ignition patterns and fire behavior trends are more relevant at a smaller burn unit scale. Seasonal vegetation curing and weather observations use daily average weather, whereas a prescribed burn event uses hourly weather. It is essential to consider the space and time scales needed for you intended observations. This will become more clear as you move through the chapters, but I wanted to highlight the importance of reflecting on the scale of observations before digging into the detailed observation approaches and exercises.

General Formats & Layouts

It can be disconcerting when you open a blank journal. I recommend formatting the first one to two pages of your journal as an index to better track and reflect back on what was collected in your journal. This requires page numbering, whether pre-printed or added manually.

Journal layouts are typically self-designed rather than built into the notebook. Layout refers to how notes, sketches, and tables are arranged on the page. I frequently use two-page spreads for all journaling approaches. With a two-page spread the words, numbers, and pictures across both pages are organized as an overall organization of related information and observations.

It is also helpful to consider any references, keys, and codes you may want to use and pre-create relevant legends and or coding systems. You can start with my examples in the book or develop your own. Either way, I recommend some thoughts about consistency in recording key contextual information such as date, time, location, and weather. From a format perspective, this can mean where you put the location, date, and time on the page and how you'd like to organize your information. While formats may vary by journaling approach, maintaining internal consistency improves later review and recall.

To create flow and meaning to the pages and layouts in your journal, I recommend using clear headings with larger or bolded lettering to make entries easier to locate. Over time, recurring headings can inform how you structure your journal index.

Formats for Different Approaches

I find it very helpful to use certain formats and layouts for the three approaches that I have described. There are formats and journaling techniques that work better for story, study, and situational awareness and can be organized and designed before you go into the field for observations. This process helps prompt you on what you need to notice and record and frame how you will describe observations.

Story Layouts—observations that convey relationships

- Sequence formats with images of people and activities including descriptions throughout the burn event including quotations and thoughts from participants or the emotional responses you have relating to the place and activities.
- Before-and-after comparisons sketches and or photos of changes to the landscape/vegetation or tied to the experience.
- Story maps that connect the experiences of people, place, and fire and emphasize the experience and relationships within the landscape
- Journal pages combining sketches, notes, and reflections that tie closely to the human relationship with fire. This can include poetry, metaphors, and creative expressions relating to the experience.

Study Layouts—-observations of key weather, fuels, terrain, fire behavior, smoke, and fire effects variables and comparisons

- Hourly data tables, emphasizing weather and other related observations that at least include date, time, and location that can be compared with hourly weather.
- Maps with numbered observation locations, emphasizing elements such as major terrain features (shaded relief maps or isomer block landscape sketches), primary vegetation categories (fuel models), or locations of burn activity locations and patterns.
- Sketches, diagrams, and photographs, with notes, to compare different fuels/vegetation concentrations, slopes, aspects, fire behavior, smoke, and fire effects conditions .
- Multi-panel layouts organized by time, location, or variable.

Situational Awareness Layouts—observations that connect environmental cues with recognition and response

- Observation checklists and environmental cue trackers emphasizing multi-sensory observations with a strong emphasis on wind speed and direction changes and changing moisture conditions in cured grasses and fallen leaves/needles.

When journaling about sensory experiences, consider adding simple visuals that help communicate your observations at a deeper level. In my example below, I included a small cartoon-style sketch of myself inhaling the scents around me. I use colors to represent different scents, both in a small legend and as swirling lines surrounding the sketch. These drawings are not essential, but they can be a powerful addition to your journal by helping convey experiences.

- Maps highlighting lookouts, hazards, escape routes, and safety zones.
- Burn parameter and associated environmental cues tracking sheets, such as the feel and sound differences when touch, bending, and or walking through curved grasses during low, optimal, and high burn parameter conditions.
- Quick-reference pages organized around recognition, interpretation, and response to environmental cues.

The images on the next page are examples of how the format and framing for Story and Study visuals might be approached differently.

Visual Formats

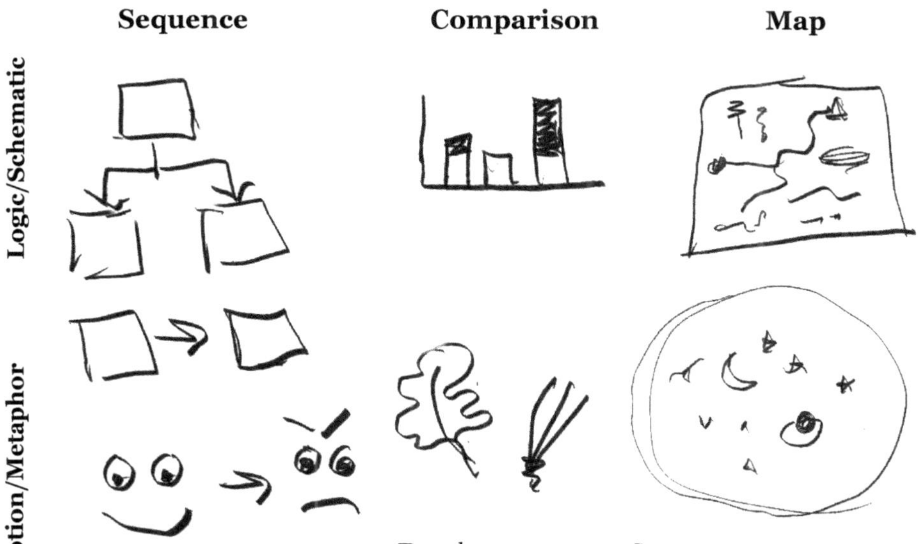

Even simple sketches can have story and artistic meaning.

Silhouettes used as more anonymous observer and participant

Journaling Supplies

Journal Size & Paper

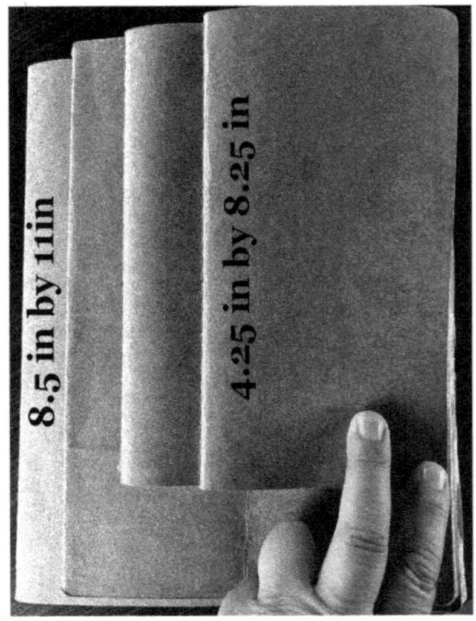

Selecting journal size is often one of the most difficult decisions. A journal that is too large becomes cumbersome in the field, while one that is too small limits data organization. I recommend a minimum size of 4.25 × 8.25 inches, with larger formats up to 8.5 × 11 inches for more extensive tracking. The commonly used A5 size (approximately 5.8 × 8.3 inches) offers a practical balance.

Cover material is another consideration. Softcovers are often more cost-effective and lighter to carry, while hardcovers offer durability. Refill-style journals can reduce costs further and allow flexible cover options.

Paper type significantly affects usability. Plain, graph, and dot-grid formats are all viable, but I strongly recommend dot-grid paper for its versatility across sketching, mapping, and data tracking. Paper weight is also important; a minimum of 100 gsm is recommended, with 120 gsm. A balance must be struck between minimalism and expressive capability.

Core Supplies

Taken on the Burn

- Dot-grid journal for ease of sketching and data tables and graphs
- Mechanical pencil with eraser, ink pen and whiteout for mistake correction
- A small number of highlighter pens (I use an orange and blue highlighter for tracking key weather parameter categories (below and above desired/optimal burn parameters)

In Vehicle or At Base Camp (use before and after the burn)

- Clear overlay material (vellum, tracing paper, or large clear sticky note sheets)

- Graphite transfer paper for tracing burn unit maps
- Scissors and double-sided tape for cutting and sticking maps and or other tables and trackers into the journal
- White gel or acrylic pen for accentuating clouds and or smoke visualization on map overlays and or sketches

Optional Journaling Additions

- Compact watercolor set
- Water brush
- Cotton cloth or wristband for brush cleaning
- Small colored-pencil set
- Handheld mini photo printer
- Page clip
- Field tote bag
- Map stencils (state boundaries)
- Stickers and Washi tape

Field Equipment

If following formal forestry or fire-monitoring protocols, additional equipment may be required. For general journaling purposes, useful tools include:

- Measuring tape for recording tree diameters and short distance measurements
- Small ruler for duff and soil depth measurements
- Weather instrument such as a Kestrel or belt weather kit
- Smartphone with relevant apps (Avenza, Google Earth, Wildland Toolkit, weather, calculator, unit conversion), and compass
- Compass, if smart phone does not have that function or app.
- Camera, if you do not carry a smartphone with camera options

Burn Framework

Before the first line of flame is ever drawn across a landscape, intention is already present. Prescribed fire begins not with ignition, but with purpose—a quiet convergence of stewardship, science, and tradition. Across generations, people have shaped ecosystems through deliberate burning, guided by cultural knowledge, ecological responsibility, and the practical need to care for land. Today's prescribed fire programs continue this lineage, blending Indigenous and cultural fire practices with contemporary fire science and land-management goals.

To journal effectively about good fire, we must first understand this framework of intention. Observations gain meaning when they are connected to why a burn is occurring and what outcomes are desired. Journaling becomes more than a record of events; it becomes a reflection of stewardship in action—a way of witnessing how human decisions shape ecological processes in real time.

This chapter introduces the foundational elements that guide good fire burning: overarching goals and specific burn objectives. Together, these elements form the conceptual and operational framework upon which all subsequent observations are built.

Burn Goals & Objectives

Most burns have overarching goals and specific objectives. It is common to see goals such as reducing hazardous fuels and improving community protection, with objectives like reducing surface fuel loading, increasing spacing between shrubs and trees, reducing ladder fuels, or creating firebreaks around homes and infrastructure. Likewise, a habitat enhancement goal may include objectives related to vegetation structure, species composition, or the creation of specific habitat conditions.

Understanding this distinction can help focus your observations. Even if you do not have access to a formal burn plan, asking a few simple questions can provide valuable journaling direction:

- What are we trying to accomplish?
- Why is this area being burned?
- What conditions are desired?
- How will success be evaluated?

While burn goals are often similar across prescribed and cultural burning, the objectives and indicators used to guide decisions may differ.

Many prescribed fire programs rely on standardized, measurable objectives. Weather parameters, fuel conditions, flame lengths, rates of spread, vegetation measurements, and other quantitative indicators provide a consistent framework that can be applied across many landscapes by practitioners with varying levels of familiarity with a particular place. These approaches support repeatable outcomes, communication, training, and operational decision-making.

It is my understanding that what distinguishes a cultural burn is that it is planned and led by Indigenous cultural fire practitioners. Cultural burning draws upon long-term relationships with place and observations accumulated across generations. Ecological conditions, seasonal timing, plant and animal life cycles, cultural resources, and other place-based indicators may help guide when, where, and how fire is applied. These indicators can reflect relationships that are difficult to capture through a single measurement but are nonetheless rooted in careful observation and experience.

In practice, these approaches frequently overlap. Prescribed burns may incorporate ecological and cultural objectives, while cultural burns may also use weather forecasts, fuel conditions, and other quantitative measures. Rather than viewing them as separate systems, journalers may find it more useful to recognize them as complementary ways of observing and understanding fire on the landscape.

For journaling, this distinction is important because objectives influence what we choose to notice. Quantitative objectives often encourage measurements, comparisons, and tracking changing conditions. Ecological and relational objectives may encourage observations of seasonal timing, species interactions, cultural values, and relationships between people and place. Most burns involve a blend of both perspectives, providing multiple ways to understand intentions, outcomes, and fire effects.

Many burn objectives are organized around desired fire intensity and specific ranges of burning conditions (burn parameters), such as temperature and relative humidity. Other objectives may be linked to ecological timing and seasonal conditions. For example, the emergence of ground-nesting bees may indicate that soil temperatures have warmed, plants are flowering, or grasses have reached stages of curing associated with desired burn conditions.

Additional considerations—such as smoke impacts on nearby communities, cultural values, and operational constraints—also shape burn prescriptions. Even when higher fire intensities are desired, prescribed burning conditions remain below thresholds associated with Red Flag fire weather conditions (NWCG, various dates).

Burn Plans

In many cases, prescribed burns are supported by an official burn plan that outlines the burn area, goals, objectives, and the burn parameters under which the burn can occur. Ideally, you can review the burn plan or request a summary of the burn objectives and prescription. If this is not possible, generalized prescription criteria can still provide a useful framework for journaling observations.

Burn plans vary widely, but many include goals related to:

- Hazardous fuels reduction and community protection
- Habitat enhancement and ecosystem restoration
- Cultural stewardship and resource objectives
- Training and workforce development
- Smoke management
- Invasive species control
- Forest and rangeland health

These broad goals provide context for understanding why a burn is taking place. The objectives, prescription parameters, and indicators associated with those goals often provide the most useful guidance for journaling observations. Throughout this book, I share examples of burn objectives, indicators, and prescription parameters as they relate to observational practices.

Observing Ecological Relationships

The indicators shown on the following pages are not intended to serve as a prescription or decision-making guide. Rather, they illustrate how seasonal patterns, weather, plant and animal activity, and fire behavior can be observed as interconnected parts of a living landscape. Their value lies not in predicting a single "right" time to burn, but in helping us understand what ecological relationships are present when a burn occurs and how the timing of fire may influence those relationships.

While measurements help define burn windows and desired fire behavior, they do not always convey the broader seasonal context created by accumulating heat, changing moisture conditions, and the progression of ecological processes across a landscape. Measures such as Growing Degree Days (GDD) and Vapor Pressure Deficit (VPD) help describe these larger environmental patterns by linking temperature, moisture, and seasonal change, but they can be difficult to visualize or relate to everyday observations.

The observational approach presented here draws from ecological science, decades of experience in fire and natural resource management, nature journaling, long-term phenological observation, and conversations with Indigenous cultural fire practitioners that have broadened my understanding of stewardship, reciprocity, and seasonal ecological relationships (Anderson, 2005; Kimmerer, 2013; Long et al., 2023).

Use the chart that follows as a starting point for identifying observations that may be meaningful where you live, work, and journal. Over time, these observations can help build connections among quantitative measurements, ecological conditions, and your own relationship with place.

As you study the chart, consider:

- Which indicators are familiar in your local landscape?
- Which indicators might help explain why a burn is planned at a particular time?
- How might ecological observations complement weather forecasts, fuel measurements, and burn objectives?
- What additional indicators could you add for your own vegetation communities and places?

Ecological/Relational Burn Indicators

Indicative Seasonal VPD (kPa)	Late Winter 0.2-0.5 Limited light/heat	Early Spring 0.3-0.8 Increasing light/heat	Late Spring 0.5-1.2 Accumulating light/heat
Plants	• Leaf buds swelling • Willow catkins	• Lupine and poppies begin flowering • Leaves unfold • Native bulbs bloom	• Peak bloom • Shrubs in bloom • Grass rapid growth
Insects	• Moths and beetles active in warmer conditions	• Native bees emerge	• Butterflies peak • Dragonflies emerge • Many pollinators active
Birds	• Early song territories established at lower elevations	• Spring migrants begin arriving • Breeding and nesting begins	• Peak Nesting
Amphibians		• Breeding begins • Eggs laid	• Adults on land, seek moist refugia
Mammals	• Bats in deep torpor • Deer in overwintering habitat	• Fawns born (varies by region)	• Pregnant bats seek maternity roost trees • Bat pups born
Potential Fire Intensity (Relative)	 Very Low Intensity	 Low Intensity	 Low to Moderate Intensity

Ecological/Relational Burn Indicators

Early Summer	Midsummer	Late Summer	Early Fall
0.8-1.6 Increasing heat/drying	1.2-2.2 + Long day, hot/dry	0.8-1.6 Decrease heat/drying	0.3-0.8 Cooler/Moister

- Grasses cure and seeds form
- Medicinal plants at peak

- Grass seed heads emerge
- Wild rose blooms

- Seeds mature/drop
- Tarweed bloom
- Fruits ripen

- Seeds drop
- Perennials store energy
- Moss & lichen active

- Grasshoppers increase
- Dragonflies abundant
- Bees & butterflies active

- Crickets & cicadas call
- Insect activity high

- Beetles active
- Predatory insects increase

- Moths active
- Cooler nights
- Insect activity declines

- Peak Nesting
- Some hatchlings and nestlings

- Some late nests
- Fledglings
- Molt begins

- Foraging increases and migration begins

- Adults on land, seek moist refugia

- Activity reduced, seek shade, cover, moisture, estivate

- Activity increases with cooler/wetter conditions
- Overwintering begins

- Bat pups nursing/ grow. Adults active foraging
- Fawns localized and stay close to cover

- Young bats indecent. All actively foraging
- Fawns more mobile and deer seek green areas

- Bat mating season and hibernation prep
- Deer rutting season

Moderate to High Intensity

High to Very High Intensity

Moderate to High Intensity

Moderate to Low Intensity

Interpreting Ecological Timing

The seasonal indicator chart is not intended to identify a single "best" time to burn. Instead, it illustrates how ecological indicators help us understand the living relationships present before, during, and after a fire. By observing plants, insects, birds, amphibians, mammals, weather, and fuels together, we begin to recognize how seasonal timing relates to burn objectives and influences fire effects.

Fire has long been part of many ecosystems, and Indigenous peoples have also shaped many landscapes through carefully timed cultural burning that supports reciprocal relationships with foods, medicines, basketry materials, wildlife habitat, and the long-term health of the land. While this book does not teach cultural burning practices, it encourages observing the ecological relationships that have long informed stewardship across many landscapes.

Rather than viewing indicators as a checklist, consider what they reveal about the living landscape and how those relationships connect to the goals and timing of the burn.

Late Winter

Plant indicator: Dormant shrubs and tightly closed buds.

Animal indicator: Ground-nesting bees remain below ground.

Timing & Stewardship: Burns may be planned to encourage vigorous shrub resprouting for weaving materials, maintain woodland structure, or reduce accumulated surface fuels. Dormant vegetation and inactive pollinators help place these objectives within their seasonal context.

Early Spring

Plant indicator: Bud burst and emerging shrub shoots.

Animal indicator: Territorial bird songs increase as nesting territories are established.

Timing & Stewardship: Burning as shrubs emerge from dormancy may influence the quality and form of new growth valued for basketry, while increasing bird activity reminds us that seasonal ecological relationships are changing rapidly.

Late Spring–Early Summer

Plant indicator: Flowering plants and abundant pollinator resources.

Animal indicator: Fawns become increasingly mobile, and many fledglings begin leaving nests.

Timing & Stewardship: Some grassland burns may be planned after young wildlife are capable of moving away from fire while also maintaining open grasslands, improving forage quality, or supporting flowering plants and pollinator habitat.

Summer

Plant indicator: Curing grasses and increasing plant moisture stress.

Animal indicator: Butterflies, bees, and many insects remain active.

Timing & Stewardship: Summer burns may support objectives such as invasive plant control, grassland restoration, or maintaining fire-adapted plant communities. Active insects and drying vegetation help place these objectives within the seasonal landscape.

Fall

Plant indicator: Ripening acorns, seed production, and the return of dormancy.

Animal indicator: Wildlife concentrates around seasonal foods while many birds migrate or shift into winter habitats.

Timing & Stewardship: Autumn burning may support oak woodland stewardship, improve access for gathering foods, enhance wildlife habitat, and prepare landscapes for the next growing season. Many plants and animals are also entering less vulnerable life stages.

Across Seasons

No single visit captures the complete story of a prescribed or cultural fire. Some ecological indicators appear only during brief seasonal windows, while others emerge months after the burn. Returning to the same Observation Point (OP) throughout the seasons often reveals relationships that would otherwise remain hidden.

Compare site with ecological observations such as flowering, insect activity, bird behavior, and wildlife use. Together they reveal not only what fire changed, but how the living landscape continues responding through time.

Operational Plans & Participants

The dawn air on a burn day carries a distinct, charged stillness, broken only by coffee chatter, equipment checks, and the ritual cadence of the morning briefing. As participants gather, operational plans translate the broader goals and objectives of the burn into the actions of the day. Weather forecasts, communication plans, assignments, safety considerations, and expected fire behavior begin to shape how the operation will unfold. By collecting selected information from briefing discussions and operational materials, you begin to assemble the contextual framework for your observations.

The information shared during briefings often extends beyond what is documented in operational plans, like an Incident Action Plan (IAP). Discussions provide valuable insights into anticipated challenges, decision-making considerations, local knowledge, partnerships, and operational priorities. For journalers, these conversations help connect the broader intentions of the burn to the specific activities taking place on the landscape.

As participants move into their assignments and ignition begins, opportunities for extended discussion often become limited. The briefing may be one of the best opportunities to gather foundational information about the day's operations, participant roles, weather conditions, and anticipated fire behavior. These notes can provide valuable context for later Story, Study, and Situational Awareness observations.

Key elements to consider recording in your journal include:

- Goals, objectives, or operational priorities
- Agendas, timelines, or schedules
- Situational awareness information
- Participant assignments and leadership roles
- Weather forecasts and expected fire behavior
- Burn-unit logistics and operational considerations
- Quotations, observations, or insights shared during discussions
- Partnerships, organizations, or communities involved in the burn

Preparing a simple journal layout before the briefing helps ensure that important information is captured efficiently. Pre-structured pages act as observational prompts, allowing you to focus on listening, witnessing, and documenting rather than designing layouts in the moment.

Briefing & Conversation Journaling

Preparing a simple journal layout before the briefing helps ensure that important information is captured efficiently. Pre-structured pages act as observational prompts, allowing you to focus on listening, witnessing, and documenting rather than designing layouts in the moment.

In my example, I lightly sketched several areas that could be refined during or after the briefing. The layout functions as a visual hub, with a central theme surrounded by related subjects. Use this format to organize goals, partnerships, participant roles, weather information, and other contextual details. Take notes and photographs throughout the event, then return later to expand and connect observations.

While operational information helps explain what is happening, observations of people often reveal how the burn is being carried out. A second layout option focuses on participants and key discussion topics. This format can be used during briefings or at any point throughout the burn event.

Create simple human figure symbols as shown in my example, or experiment with full-body sketches and portraits if time allows. When using simplified figures, include one or two distinguishing features. Sketches should remain respectful and observational rather than caricatured. Subtle visual cues—such as hat shape, posture, or equipment—can help differentiate individuals without requiring precise likenesses.

Begin by recording names, quotations, comments, or observations. Topic headings can anchor related notes, with names and insights arranged around them. Borders, frames, or simple graphic elements can help visually connect people with shared ideas, responsibilities, or discussions.

When documenting extensive notes on a single topic, consider grouping information using distinct shapes, frames, or color highlights. For example, one theme might be enclosed in a solid border while another uses dashed lines or a different highlight color.

Be mindful when recording personal stories, culturally sensitive information, or Indigenous knowledge shared during the burn. Sensitive observations may be recorded in pencil and revisited later with participants if questions arise regarding inclusion. Often the concern is not personal documentation, but how recorded information may eventually be shared, interpreted, or attributed.

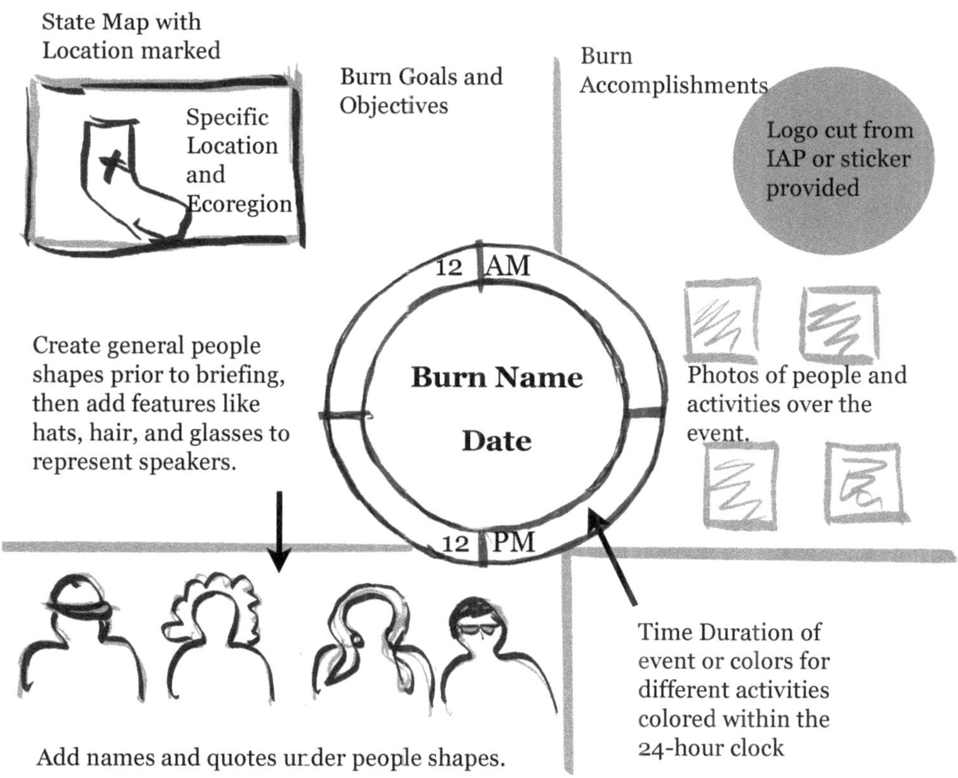

Story Journal Examples

As a Story approach, I like to summarize the event. In the example above, I included the workshop and burn agenda, some of the people I worked with, and a few highlights from the event. The example on the right includes a map showing the event location, a photo of the training cadre, a group photo of the burn participants, and a list of the burn partners who supported the training.

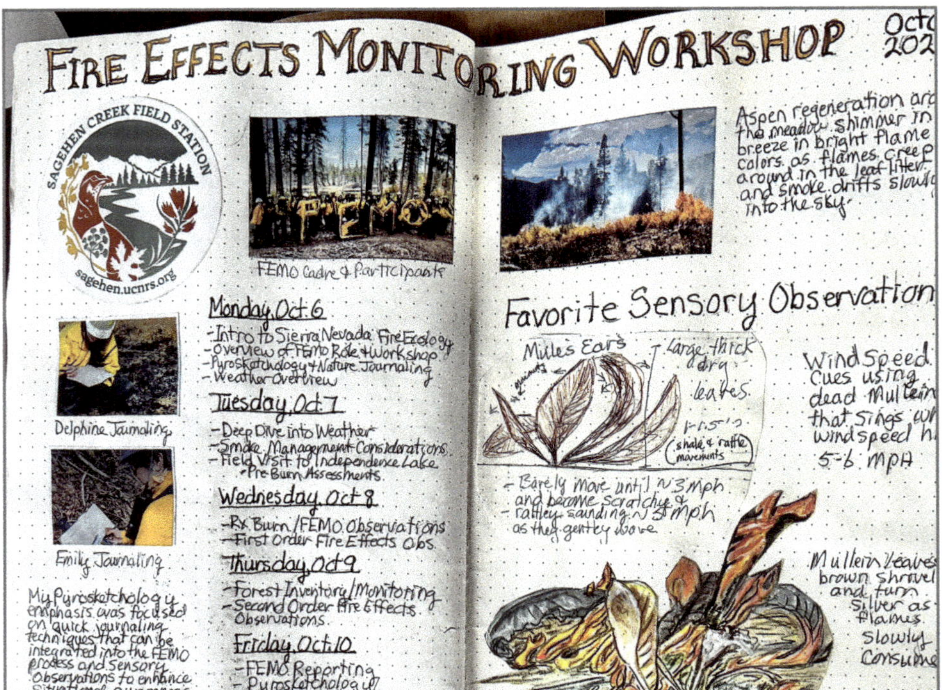

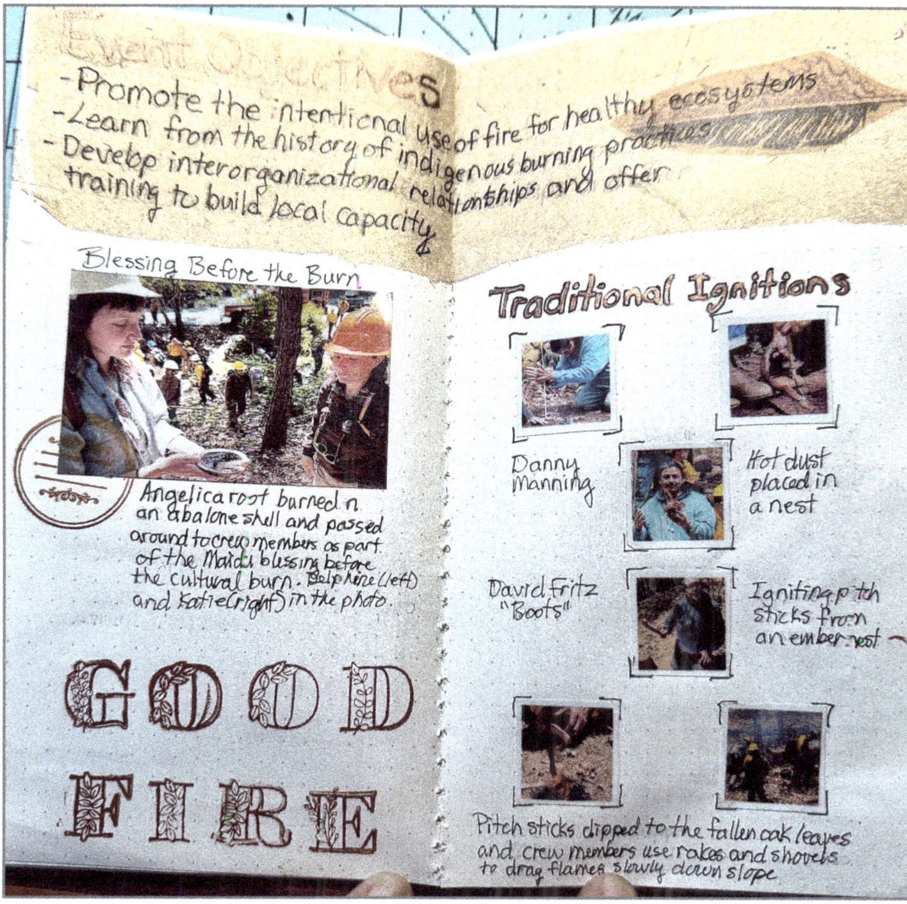

In the example above, I included the overarching burn goals along with a few photos and notes about the people and the traditional cultural fire ignition techniques we learned, which were the highlight of the event for me.

Poetry & Personal Connection

Throughout this chapter, observations have been framed through goals, objectives, operational plans, ecological indicators, and participant experiences. Poetry offers another way to translate those observations into meaning. In fire work, where decisions are guided not only by science but by intention, experience, and care, poetry can help clarify why we burn—not just how. It connects emotion to objective and purpose to place. That connection is often what drives good fire.

I was introduced to a simple but impactful poetry exercise during a workshop with Aimee Nezhukumatathil at the Wild Wonder Nature Journaling Conference. The structure below adapts that exercise for fire practitioners and observers, grounding creative expression in ecological purpose and field experience.

Begin by brainstorming responses to the following prompts. Move quickly and intuitively—this is not about perfect wording, but about noticing what arises.

- Name a goal or objective of good fire that fills you with hope or inspiration.
- Describe a specific location where you observe that objective being applied—but describe it without naming it. Let the place emerge through detail. For example: tall, wind-brushed grasses in a high mountain meadow.
- Describe the timing—not with a clock, but through your senses. What does the light look like? What is the quality of the air? For example: under a deep blue sky streaked with slow-drifting cottontail clouds.
- Describe a physical or biological transformation tied to your objective. For instance: scratchy, dry grasses yielding to ash that settles into the seedbed of the next generation.
- Add another physical or biological characteristic—perhaps something subtle or easily overlooked.
- Choose one action verb that captures what fire is doing in this moment.
- Name one form of harm that can result from fire—or from the absence of fire.
- Describe another location within the same burn unit or ecosystem where this good fire is envisioned or already unfolding.

Once you have gathered your notes, arrange them into something resembling a poem. Let the structure emerge naturally—short lines, fragments, or complete sentences are all valid. Focus on rhythm, contrast, and clarity of image rather than strict poetic form.

Pair your poem with sketches or photographs that help anchor it in lived experience. Over time, these entries can become a record not only of fire effects, but of your evolving relationship with fire—how your understanding, values, and observations shift across seasons and landscapes.

Bringing It Together

The tools and approaches introduced in this chapter are not intended to create more work or turn every burn into a detailed study. Rather, they provide a framework for choosing what to observe and how to organize those observations in ways that are meaningful to you.

Some days your journal may focus on operational details, weather forecasts, and changing fire behavior. Other days you may be drawn toward ecological indicators, conversations, partnerships, or the stories unfolding across a landscape. At times, a quick situational awareness note may be all that is practical. At other times, a sketch, map, poem, or reflective entry may capture a deeper experience of fire and place.

The Story, Study, and Situational Awareness approaches are not separate methods so much as complementary ways of engaging with the same burn. Together they encourage us to observe not only what fire is doing, but why it is occurring, how it is being applied, and what relationships it reveals among people, place, weather, fuels, and ecology.

As you move into the chapters that follow, use these approaches as flexible guides rather than rigid structures. Build your own combinations of observations, sketches, notes, and reflections. Over time, your journal will become more than a record of prescribed or cultural burning. It will become a record of your evolving understanding of good fire and your relationship with the landscapes and communities it serves.

Landform & Place

A landscape is more than a collection of elevations, slopes, and compass directions. It is a living expression of weather, water, sunlight, vegetation, and time. The shape of the land influences where clouds gather, where moisture lingers, where plants thrive, and how fire moves across a landscape. Over generations, people have learned to recognize these patterns through direct experience, observation, and relationship with place.

Fire practitioners often use the term *topography* to describe the physical features of a landscape—its elevation, slope, and aspect. These features are important because they influence weather, vegetation, fuels, and fire behavior. Yet landscapes are more than topographic measurements. They are places experienced through shifting sunlight and shadow, seasonal changes in moisture, ecological communities, cultural connections, and personal memories. The same hillside can be measured on a map, observed through the characteristics of its plants, or understood through the stories people associate with it.

In this chapter, I explore how landforms help shape the character of a place and how journaling can help us observe, document, and better understand those relationships. Landscapes can be understood through many lenses. Ecological indicators such as vegetation characteristics, seasonal patterns, wildlife activity, moisture conditions, and cloud development help reveal the environmental gradients associated with elevation, aspect, and landform. Measurements such as elevation, aspect, slope, and geographic location provide standardized ways to document and compare observations. Maps, sketches, and spatial journaling techniques help connect these observations across scales, from individual observation points to entire watersheds and burn units. Although this chapter focuses on landform and place, some observations naturally overlap with fuels, vegetation, weather, and fire behavior because these elements are experienced together in the landscape. Together, these approaches create a foundation for understanding place while introducing relationships that will be revisited and expanded throughout the chapters that follow.

As with many Pyrosketchology practices, I encourage you to work across multiple scales of observation. At one moment, you may be studying an entire mountain range or watershed. At another, you may be focused on the angle of a single slope, the shape of a tree, or the way afternoon sunlight warms a patch of grass. Over time, these observations begin to form an internal landscape model—a mental framework that helps you anticipate changes in weather, fuels, and fire behavior before they occur.

Burn Unit & Landscape Context

The burn unit provides the physical framework within which prescribed or cultural fire activities occur. It is the portion of a landscape where burn objectives, ecological conditions, operational considerations, and fire behavior intersect. Understanding the burn unit helps define not only where fire will occur, but also where observations can be focused.

Burn units are typically bounded by natural or constructed firebreaks/control lines. These may include roads, trails, rock outcrops, water bodies, wet lines, recently burned areas, or hand-constructed fireline. Such features help contain fire within the intended area and influence ignition patterns, holding operations, and observed fire behavior.

While burn-unit boundaries are important operational features, journalers should also consider the broader landscape surrounding the burn. Fire behavior is influenced by factors that often extend beyond the burn-unit perimeter, including ridges, valleys, drainages, vegetation transitions, and weather patterns.

Observing these larger landforms provides context for understanding why a burn unit was selected and how fire may interact with the surrounding environment.

One useful journaling exercise is to begin by locating the burn unit within the larger landscape. Notice where ridges, valleys, streams, roads, and vegetation transitions occur. Consider how these features influence access, containment, ignition strategies, smoke movement, wildlife habitat, and potential fire behavior. These observations help transform a mapped boundary into a more complete understanding of place. See my journal page example on the next page.

Throughout this chapter, I encourage you to move between the burn unit and the larger landscape. Understanding both scales helps build a stronger sense of place and creates a foundation for interpreting weather influences, ecological relationships, operational activities, and fire behavior.

Indicators of Elevation & Aspect

Elevation and aspect are often recorded as measurements on a map, but they can also be understood through the characteristics of a landscape itself. Together they influence sunlight, temperature, moisture, wind exposure, growing season length, vegetation communities, fuel conditions, and fire behavior. These environmental gradients help shape the ecological character of a place and influence how fire interacts with the landscape.

Changes in elevation often create broad transitions in weather, vegetation, and fire regimes. Aspect creates similar patterns at a finer scale. A north-facing slope may experience environmental conditions more similar to a higher-elevation landscape, while a south-facing slope may resemble conditions found at lower elevations. In both cases, plants, animals, fuels, and seasonal patterns reflect the combined influences of sunlight, moisture, temperature, and landform.

Long before topographic maps, GPS devices, and digital terrain models existed, people learned to recognize these relationships through direct observation and experience. Changes in vegetation, wildlife activity, weather patterns, and seasonal indicators provided information about environmental conditions that could influence travel, food gathering, cultural practices, and the timing of fire. Many of these observations remain useful today, helping us connect measurements and maps with the lived experience of a place.

These patterns become visible through observation. In many landscapes, valley bottoms and lower elevations often support taller vegetation, broader leaves, deeper soils, and greater moisture availability. Riparian areas may contain dense plant communities adapted to abundant water and longer growing seasons. Trees frequently develop larger crowns, denser canopies, and greater overall height where moisture is more abundant and growing seasons are longer. Seeds and fruits may also be larger or more numerous, reflecting the greater energy available for growth and reproduction.

Middle elevations often contain a blend of characteristics from both lower and higher landscapes. Vegetation communities become more varied as temperature, moisture, and exposure interact across slopes and drainages. Trees may be moderately sized and widely distributed across the landscape, while leaves, cones, seeds, and fruits reflect adaptations to both seasonal moisture limitations and periodic environmental stress. These transition zones often support high ecological diversity and provide opportunities to observe how changing environmental conditions influence plant communities.

At higher elevations, temperatures generally become cooler, growing seasons become shorter, and environmental conditions become more variable. Trees and shrubs are typically shorter, more widely spaced, and increasingly influenced by wind, snow, and cold temperatures. Near upper elevation limits, trees may appear twisted by persistent winds, bent by heavy snow loads, or stunted by harsh growing conditions. Leaves often become smaller, thicker, waxier, or needle-like as plants adapt to greater exposure and shorter growing seasons. Cones, seeds, and fruits may also differ from those found at lower elevations, reflecting the challenges of shorter growing seasons, colder temperatures, stronger winds, and seasonal snow cover. Observing their size, shape, protective structures, and methods of dispersal can provide additional clues about how plants are adapted to a particular environment and the conditions that shape a place.

Similar patterns can often be observed across aspects. South- and west-facing slopes typically receive more direct solar radiation, resulting in warmer temperatures, drier fuels, and greater moisture stress. These conditions often favor more drought-tolerant vegetation, earlier curing grasses, and more open plant communities. North- and east-facing slopes generally retain moisture longer, experience less intense solar heating, and often support denser vegetation, larger leaves, and later seasonal development. Together, these observations provide clues about how sunlight, moisture, temperature, and landform interact to shape a place.

When journaling, consider observing:

- Tree height, shape, spacing, and branching patterns
- Leaf size, thickness, texture, and seasonal color changes
- Cone, seed, fruit, or flower characteristics
- Transitions in dominant vegetation communities
- Differences in vegetation density between slopes and valley bottoms
- Flowering, fruiting, and curing patterns
- Wildlife activity and indicator species
- Snow persistence, frost pockets, or seasonal moisture patterns
- Fuel continuity, fuel bed depth, and vegetation structure
- Areas of prolonged sun exposure and shade
- Differences between ridges, slopes, benches, drainages, and valley bottoms

Together, these observations help reveal how elevation and aspect influence ecological conditions, fuel characteristics, and fire environments.

Elevation, Aspect & Location Data

While indicators provide valuable context, recording elevation, aspect, and location data remains important for many study-oriented journaling activities. These measurements help support comparisons among observation sites and are commonly used in fire-effects monitoring, fire-weather observations, fuel-moisture calculations, and other fire-management applications.

Latitude & Longitude

Latitude and longitude (Lat & Long) provide a standardized method for identifying observation locations. These coordinates are commonly recorded in Fire Effects Monitor (FEMO) reports, operational maps, and other fire documentation. Recording coordinates allows observation points to be revisited, compared over time, and integrated with maps and monitoring datasets.

Latitude lines run east-west and measure distance north or south of the equator. Longitude lines run north-south and measure distance east or west of the Prime Meridian. Coordinates may be obtained from GPS devices, smartphones, digital mapping applications such as Avenza Maps or Google Earth, or operational materials provided during a burn.

Elevation

Elevation influences temperature, humidity, precipitation, fuel moisture, vegetation communities, and fire behavior. Many fire-weather references and fuel-moisture calculations require elevation data to determine which charts, corrections, or reference tables should be used.

General elevation categories commonly used in fire-weather and fuel-moisture references include:

- 0–500 ft — Valley / Coastal Plains
- 501–1,900 ft — Low Hills
- 1,901–3,900 ft — Foothills (Lower Montane)
- 3,901–6,100 ft — Lower Mountains (Montane)
- 6,101–8,500 ft — Upper Mountains (Upper Montane)
- 8,501–11,000 ft — Subalpine / Alpine

Although vegetation communities vary among regions, these categories provide a useful framework when applying fire-weather and fuel-moisture references. When possible, pair elevation measurements with observations of vegetation, weather, and ecological conditions to build a richer understanding of place.

Aspect

Aspect refers to the compass direction a slope faces and is commonly recorded as north, east, south, or west. Aspect influences solar exposure, fuel temperatures, fuel moisture, vegetation characteristics, and fire behavior.

When recording aspect, it can be helpful to think beyond compass directions and consider how sunlight interacts with the landscape throughout the day. East-facing slopes typically receive morning sunlight and may warm earlier in the day, while west-facing slopes often experience greater afternoon heating. South-facing slopes generally receive the greatest solar exposure, while north-facing slopes tend to remain cooler and retain moisture longer.

These differences influence fuel conditions, vegetation patterns, and the timing of fire behavior changes throughout the burn period. Recording aspect alongside elevation provides important context for interpreting weather, fuels, and fire observations.

The Campbell Prediction System provides another useful way of visualizing aspect by emphasizing how fuel-bed temperatures change as the angle of the sun shifts throughout the day. Observing when slopes enter sunlight, become shaded, and retain heat can help explain changes in fuel moisture and fire behavior.

When using fire-weather references, fuel-moisture guides, or Dead Fuel Moisture Content Correction charts, be sure to record both elevation and aspect at your observation point. Combining these measurements with observations of vegetation, weather, and ecological indicators helps connect quantitative data with a richer understanding of place.

Slope

The steepness of a slope is one of the easiest landform characteristics to feel and one of the most difficult to estimate accurately. We experience slope through movement, balance, effort, visibility, and perspective. Fire responds to slope as well, often spreading more rapidly uphill as flames and heat preheat fuels above them.

Before reaching for a map, clinometer, or smartphone application, I encourage you to develop an intuitive sense of slope through observation and movement. Over time, these observations help build a mental reference for recognizing how steepness influences vegetation, fuels, accessibility, and fire behavior. This intuitive understanding can improve both journaling and situational awareness while helping you interpret slope measurements and maps more effectively.

Sensing & Journaling Slope

One of my favorite ways to develop a sense of slope requires nothing more than a pen or pencil and a notepad or journal.

Hold the pen vertically at arm's length while facing a hillside. Slowly rotate the pen until it matches the angle of the slope. As you compare different hillsides, pay attention to how much the pen must lean from vertical and how that angle changes across the landscape. Repeat the exercise while observing ridges, valley walls, road cuts, and other landforms.

For journaling purposes, choose a slope that is representative of the landscape you are observing, or select two examples that represent the lower and upper range of slopes within the area. At each observation point, use the average angle suggested by the pen and slope to create a simple cross-section diagram in your journal.

I usually begin by sketching a single line that captures the overall curve and angle of the slope between two points. I often use trees, shrubs, rock outcrops, or other landscape features as the start and end points. Draw the average slope angle, unless you want to include smaller dips, benches, or rises that help characterize the landform. Add a few vegetation features and record your estimate of the slope steepness.

Later, compare your estimate to measurements obtained from a clinometer, smartphone application such as the Wildland Toolkit, a topographic map, or a digital mapping platform. Over time, this practice helps develop an intuitive sense of slope while improving your ability to recognize meaningful differences in steepness across a landscape. In later chapters, this same cross-section format can be expanded to include fuels, fire behavior, smoke movement, weather observations, and other landscape relationships.

For journaling purposes, it can be helpful to group slopes into a few simple categories:

- **0–5%** — Flat
- **6–19%** — Gentle
- **20–30%** — Moderate
- **Greater than 30%** — Steep

These categories provide a useful framework for comparing observation sites and organizing journal entries. They also help develop a visual sense of how slope influences fuels, vegetation patterns, access, and fire behavior.

> Determining whether a slope is above or below **30 percent** may be more important than calculating an exact value. This threshold is commonly used in fire-weather and fuel-moisture references, including Dead Fuel Moisture Content Correction charts. It also provides a practical situational awareness benchmark.

Slope can be estimated visually, measured with a clinometer, calculated from contour maps, or obtained from digital mapping applications. The Wildland Toolkit smartphone application provides a convenient slope calculator by allowing the phone to be placed directly on the ground surface. Taking several measurements across an observation area can help develop a representative average slope value.

Regardless of the method used, combining measurements with direct observation helps build a stronger understanding of how slope influences landform, fuels, and fire. In later chapters, I will build on this foundation by exploring how slope interacts with fuel arrangement, ignition patterns, fire spread, and fire effects.

Maps as Ways of Seeing Place

Maps can serve many purposes within a journal. Some help document observations and measurements. Others help tell the story of a place or visualize relationships that are difficult to capture through words alone. Still others help anticipate changing conditions and strengthen situational awareness. Regardless of the format, maps provide a way to organize observations across space and connect details that might otherwise appear unrelated.

In Pyrosketchology practices, maps are not intended to replace formal operational maps, GIS products, or fire-behavior models. Rather, they provide a complementary observational framework that strengthens pattern recognition, ecological understanding, and place-based awareness. The process of creating a map often becomes just as valuable as the finished product because it requires careful observation, interpretation, and reflection.

The boundaries between Story, Study, and Situational Awareness maps are often fluid. A map created to tell the story of a place may also reveal patterns useful for situational awareness or future study. Likewise, a study map documenting observations may later help explain relationships that become part of a larger story. The examples that follow are organized by their primary purpose, but many maps support multiple journaling approaches.

Before exploring specific map styles, it is worth considering when and how maps are created. Some maps are prepared before observations begin, while others emerge gradually through field notes, sketches, and reflection. In practice, most journal maps evolve over time as observations accumulate and understanding of a place deepens.

Preparing Maps & Spatial Layouts

Maps are often most effective when some preparation occurs before observations begin. Whether you are creating a Story, Study, or Situational Awareness map, a little advance planning can help you spend less time constructing layouts in the field and more time observing. The process of tracing, sketching, and organizing spatial information often reveals questions worth exploring in the field and helps transform a map from a navigational tool into an observation tool.

Whenever possible, I recommend reviewing maps, aerial imagery, or landscape references before a burn. A simplified landscape sketch created from Google Earth, a topographic map, or a burn-unit map can provide a useful foundation for later observations. Even if the final map is completed after the burn, creating a basic framework beforehand helps build familiarity with the landscape and encourages you to begin noticing important terrain features, vegetation patterns, observation points, and other areas of interest.

If a burn-unit map is available, you may wish to trace or simplify it the day before the burn or during the morning briefing. Observation points, roads, drainages, ridges, and other prominent features can be added in advance, leaving space for notes and observations collected throughout the day.

When preparation time is limited, consider reserving a journal page for a future map. During field observations, you can quickly sketch ignition patterns, smoke movement, observation locations, notable fire behavior, vegetation transitions, or other spatial relationships. These rough notes often provide enough information to create a more complete map later when time allows.

The goal is not to create a perfect map before the burn begins. Instead, think of maps as evolving observation frameworks. A few lightly sketched lines, symbols, or observation points can provide structure while leaving plenty of room for discoveries that emerge through direct experience.

Depending on the size of your journal, maps may be folded, taped into the journal, or spread across multiple pages. I often simplify or trim maps to conserve space and improve note visibility. At times, I attach a sheet of vellum or a transparent overlay along the top edge of a page. This allows ignition patterns, holding activities, smoke movement, weather observations, or other information to be recorded without obscuring the original map.

Over time, these maps can be refined, expanded, or combined with additional sketches, notes, and observations as your understanding of a place grows.

Story Maps

Story maps emphasize relationships, experiences, and the character of a place. Rather than striving for precise scale or geographic accuracy, they highlight observations that help explain how a landscape is experienced and understood. A story map may emphasize a ridgeline, a wet meadow, a stand of trees, a cultural feature, a recurring wind pattern, or another element that helps tell the story of a burn unit or landscape.

In many cases, story maps intentionally simplify or exaggerate features to emphasize relationships. A mountain may be drawn larger than scale to illustrate its influence on weather and cloud formation. A creek may become more prominent because it shapes vegetation patterns, wildlife activity, or burn operations. Vegetation communities, landmarks, and terrain features can be simplified into symbols or sketches that help communicate the character of a place rather than its exact dimensions.

Story maps are particularly useful for connecting observations across larger spatial scales. They can help illustrate how surrounding mountains influence winds and clouds, how vegetation changes across a landscape, or how cultural, ecological, and operational elements interact. Because they emphasize relationships and experience, story maps often support situational awareness by drawing attention to features and patterns that might otherwise be overlooked.

The examples that follow illustrate several approaches to story maps with different intentions for observation and storytelling. All three of these story-map styles can also be used effectively for situational awareness.

The example below uses a burn-unit boundary but takes a story approach by accentuating slope and a key terrain feature that helped me remember where I was located within the unit. A few simple lines help convey a sense of slope without being drawn to scale. The goal is not precise measurement but a sense of terrain within the unit. As a note about other details on the map, the "DP" labels reference drop points identified on the map and often marked with signs on the ground. These are used for communicating locations and meeting points.

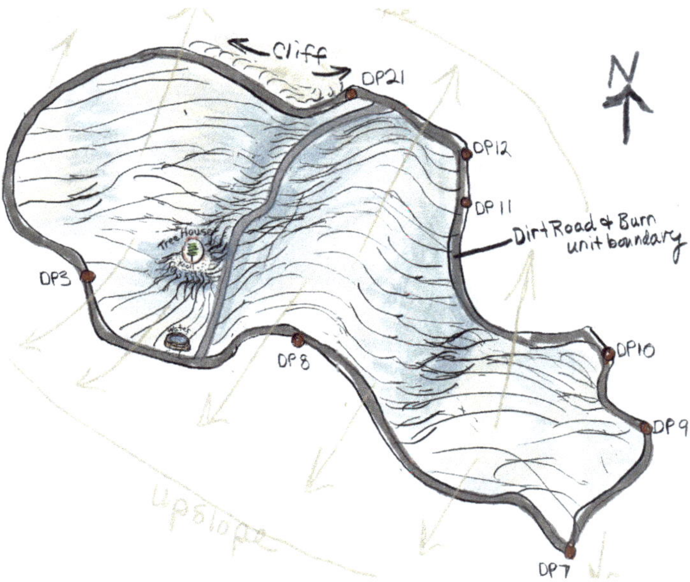

The map to the right is more of a sketch, but it is intended to show where clouds are likely to develop and the directions from which they typically arrive. This is the same location as the map above, but at a different scale. I needed a broader landscape perspective to better observe cloud movements over the mountains north and south of the burn unit. I use simple blue swirl lines to indicate common wind and cloud movement patterns.

The map below uses a story-map style with a few accentuated slope lines, terrain features (knolls), and vegetation elements (tree species and mixtures) that are not to scale but provide a better sense of what was important and observed while moving through the burn unit. The intention was not to map all terrain and vegetation features accurately across the entire unit, but to provide a general sense of what was where. I also chose to emphasize memorable experiences with short haiku-style notes in key locations. I would call this a haiku story map. It is a useful way to create a spatial story about an experience, while also supporting situational awareness.

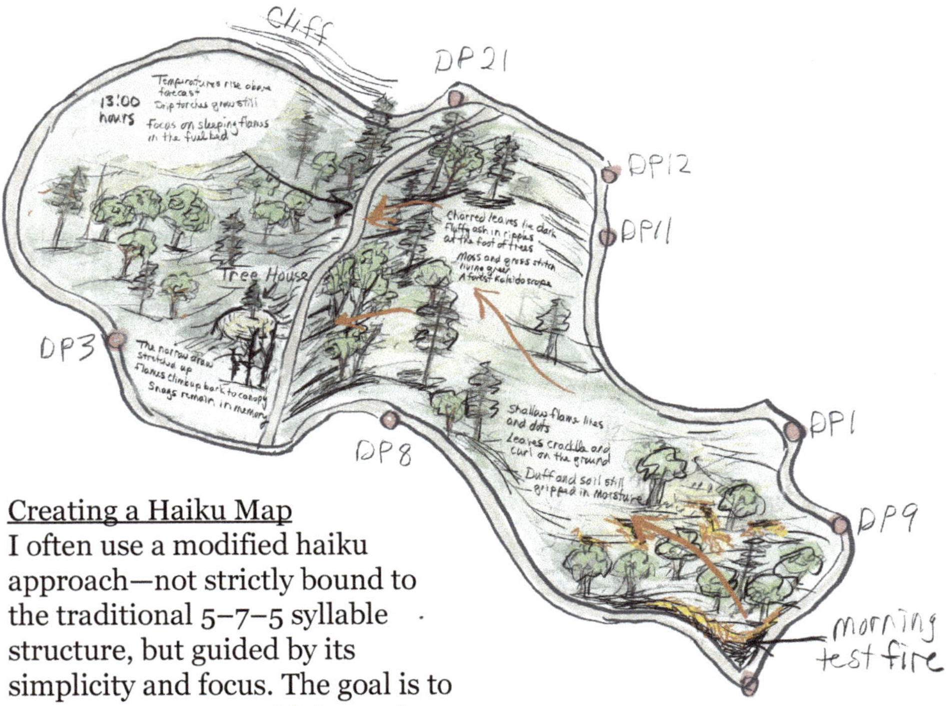

Creating a Haiku Map

I often use a modified haiku approach—not strictly bound to the traditional 5–7–5 syllable structure, but guided by its simplicity and focus. The goal is to capture a moment with just a few lines, allowing the experience of place to come through.

Some key considerations for the haiku map include:

- A word or phrase that connects to season or time
- A brief description of a fleeting moment—an action, experience, or perception
- A comparison between distinct elements—creating juxtaposition and a sense of relationship or synergy

Although story maps often appear less precise than operational or monitoring maps, they frequently reveal relationships that might otherwise be overlooked. By emphasizing what is important rather than what is exact, they help transform locations into places with meaning and context.

Study Maps

Study maps emphasize documentation, comparison, and repeatable observations. They provide a framework for organizing information collected before, during, and after a burn while helping connect observations to specific locations.

A simple burn-unit map is often sufficient. Observation points, ignition patterns, vegetation transitions, holding operations, monitoring locations, or other features can be added as observations accumulate. Recording the timing of observations at specific locations strengthens the connection between place and process and creates opportunities for later comparison.

Study maps are particularly useful when observations are tied to measurable data such as latitude and longitude, elevation, aspect, slope, weather observations, fuel conditions, or fire effects. They provide a spatial framework for organizing observations associated with the Study journaling approach and can support fire-effects monitoring, prescribed-fire documentation, and other forms of field assessment.

Although study maps often emphasize measurements and documentation, they should not become cluttered with excessive detail. The goal is to create a clear visual framework that helps organize observations and identify patterns. A simple map that is easy to use in the field is often more valuable than a highly detailed map that becomes difficult to update during active observations.

The example on the next page show an operational map from an Incident Action Plan (IAP) that was traced into my journal using carbon transfer paper. This is a quick way to get a map into your journal and add whatever amount of features and attributes that serve my purpose. You can also bring transparent paper, such as vellum or tracing paper, and tape it over the map, or use larger clear sticky-note sheets for layering visuals and notes.

Traced Burn Unit Map

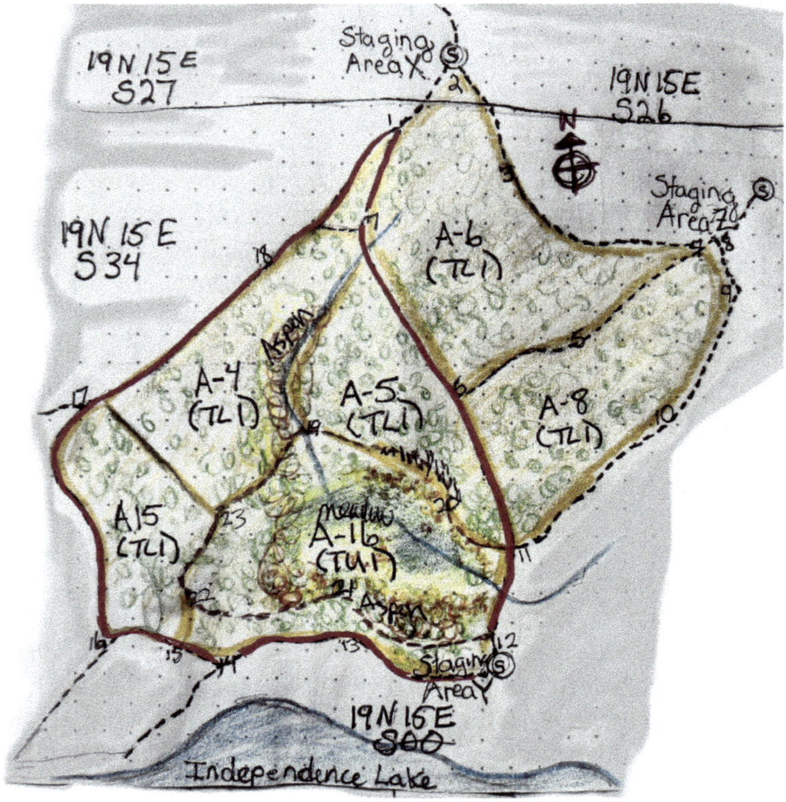

The example on the next page is a bit more complex. This type of map is often called a block landscape and provides an isometric, or three-dimensional, perspective of the land. These maps do not need to be highly detailed, but they should include enough information to convey the overall terrain and slope relationships.

I have visualized the key steps for creating a block landscape map below. I would note that these are usually larger maps with space to add terrain and vegetation elements.

Creating a Block Landscape Map

In pencil, sketch an angled square block.

In pencil, sketch outlines of features like roads, and use lines to give a sense of slope.

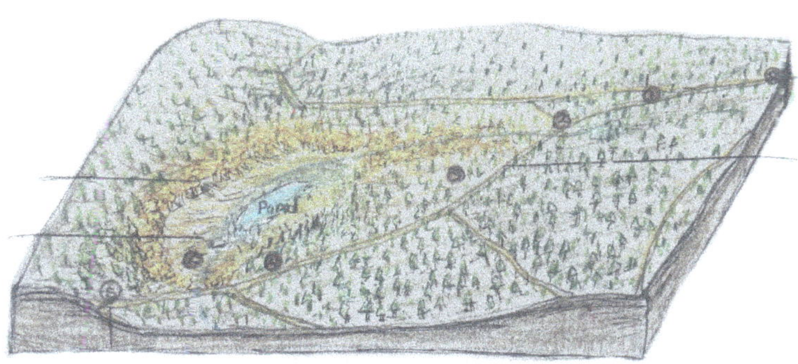

Use watercolor, colored pencils or colored pens to accentuate vegetation and other features.

Situational Awareness Maps

Situational Awareness maps emphasize where and when observations may be most important for enhancing awareness of changing burn conditions and increasing risk. These maps help observers recognize environmental cues that indicate when conditions are moving from moderate toward higher fire-intensity conditions—the times and places when heightened awareness is most valuable. For example, a map may emphasize south-facing slopes greater than 30 percent, areas where fuel-bed heating is greatest, or locations and times when fuel moisture and wind conditions approach the higher end of the burn prescription.

These maps support the ongoing process of recognizing, interpreting, and responding to cues that may indicate shifts in fire behavior potential or burn intensity.

The map to the right is a landscape-scale example focused on how diurnal terrain winds may influence the burn unit. The burn unit is identified as a point within the larger landscape, while the emphasis is placed on ridgelines, valleys, and other terrain features within the watershed. This is actually the same location shown in the earlier cloud story map. Both maps support situational awareness, but this example is more spatially accurate and therefore moves away from the story-map style.

On the next page, I share one of my favorite examples for situational awareness. It combines a simplified 3D terrain map with concepts from the Campbell Prediction System (Campbell, 2004) related to fuel-bed heating based on the aspect and time of day. By placing 24-hour clock diagrams with temperature color codes around a landscape map, it becomes possible to visualize how solar exposure and fuel-bed temperatures change across different slopes and aspects throughout the day.

The color-coded time bands would vary by season and location, but this example is generalized for the warmer times of the year. This approach provides a strong foundation for understanding daily surface-temperature fluctuations. These changing conditions influence fuel moisture, terrain-driven winds, fire behavior, and many of the sensory cues discussed throughout this book.

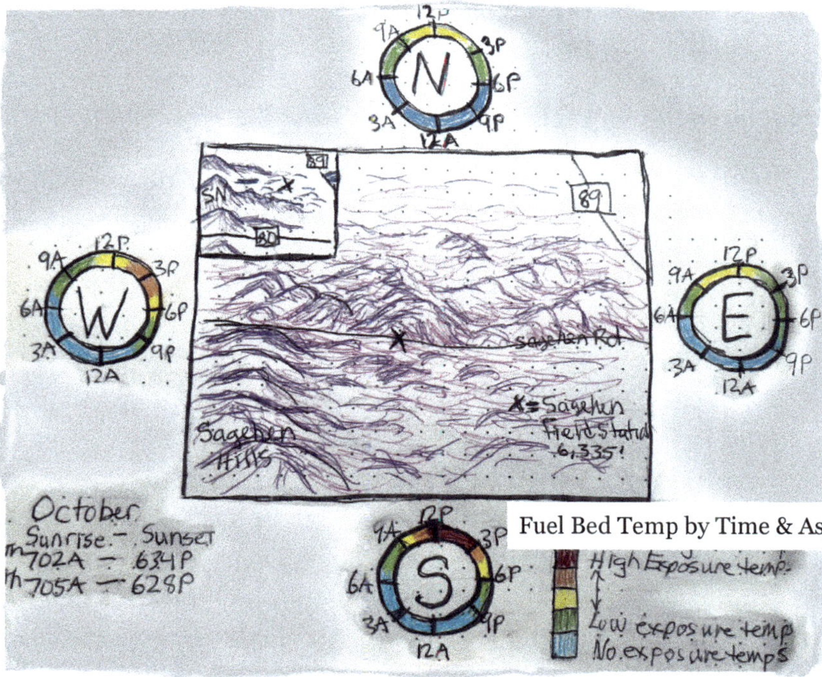

Fuel Bed Temp by Time & Aspect

Creating the Map

To create the simple 3D-style map, I used Google Earth in 3D mode as a visual reference and selected a viewing direction that best accentuated the shape of the landscape. I then captured the image and used it as a reference for sketching the larger terrain features.

I selected a single color—purple—for sketching the larger terrain features to avoid confusion with vegetation or rock characteristics. Green or brown can imply ecological conditions rather than landform. I typically begin by sketching ridgelines, then add lines descending from the ridges to convey slope and steepness. Steeper slopes often appear as straighter, more angular lines, while gentler slopes are represented with softer curves. Try not to overdo the details; focus on capturing the essential shape of the landscape. Once the major features are established, add additional lines or a light color wash to suggest shading and relief.

For the Campbell Prediction clocks, I traced around a quarter to create a consistent circle size. These circular diagrams may be arranged in a row or positioned around the map according to the four primary aspects, depending on personal preference. I then added an inner circle, divided the outer ring into broad time blocks, and created a color legend representing relative fuel-bed temperatures. The appropriate colors were then applied to the time blocks according to the legend.

After completing the map with the four aspect clocks and temperature colors, I reflect on how each slope heats differently throughout the day and how those changes influence situational awareness. You could apply the colors directly to the landscape itself, but I find that approach more confusing and difficult to interpret. This version has proven to be the clearest and most useful for my own observations. Use the following terrain alignment and diurnal wind section for reflecting on your map and what it means for changes in wind direction and speed based on terrain alignment.

Terrain Alignment & Diurnal Winds

When evaluating a landscape, it can be helpful to first look at the overall alignment of mountain ranges, ridges, valleys, and drainages. These landforms often act as channels that guide daily wind patterns. Once you recognize the dominant terrain alignment, you can begin anticipating where terrain-driven winds may be strongest, weakest, or longest lasting throughout the day.

North–South Aligned Valleys and Mountain Ranges

Landforms that run primarily north to south often experience strong differences in solar heating between east- and west-facing slopes. East-facing slopes warm first in the morning, while west-facing slopes warm later and often remain warm longer into the evening.

In these landscapes, upslope winds often begin earlier on east-facing slopes as they receive morning sunlight. During the afternoon, west-facing slopes may produce stronger upslope winds as solar heating shifts westward. Valley bottoms frequently develop up-valley winds during the day as warming draws air toward higher elevations. After sunset, cooling begins on the slopes and cold air drains downslope and down-valley toward lower elevations.

Because heating shifts from one side of the valley to the other throughout the day, wind directions and intensities may vary considerably between morning and afternoon. Situational awareness observations should consider which side of the valley is receiving direct sunlight and how that may influence local wind behavior.

East–West Aligned Valleys and Mountain Ranges

Landforms aligned primarily east to west often experience more uniform heating along valley sides because both slopes receive sunlight during portions of the day. However, the southern side of the valley generally receives more direct solar exposure throughout much of the year, while the northern side remains cooler and more shaded.

In these landscapes, upslope winds may become strongest on south-facing slopes where solar heating is greatest. Up-valley winds often develop as the entire valley warms, but the timing may be more consistent along the length of the valley than in north-south-oriented systems. During the evening, cold air drainage commonly produces downslope and down-valley winds that follow the valley corridor toward lower terrain.

Because south-facing slopes receive greater heating, these areas often experience earlier fuel drying, stronger thermal activity, and more pronounced afternoon upslope winds.

Narrow Canyons and Drainages

Narrow drainages and steep canyons often amplify terrain-driven winds regardless of their orientation. Airflow becomes concentrated as it moves through constricted terrain, producing stronger upslope, downslope, up-valley, or down-valley winds than those found on adjacent open slopes.

These areas are particularly important for situational awareness because wind speeds and directions may change more rapidly than expected. Narrow drainages can also channel smoke movement and influence fire spread patterns.

Broad Valleys and Basins

Broad valleys and basin landscapes often produce weaker terrain-driven winds because airflow is distributed across a larger area. However, large-scale valley systems may develop persistent up-valley winds during the day and strong cold-air drainage patterns at night. These winds often develop more gradually but may influence larger portions of the landscape.

Using Terrain Alignment in Journaling—When reviewing a map or landscape sketch, begin by asking a few simple questions:

- Which direction do the major valleys and drainages run?
- Which slopes receive morning sunlight and which receive afternoon sunlight?
- Where might air be funneled or concentrated?
- Which areas are likely to remain cooler, shaded, or sheltered?
- Where might upslope, downslope, up-valley, or down-valley winds be strongest?

Adding arrows to maps and landscape sketches can help visualize the observed and or expected wind patterns. When combined with aspect observations and the Campbell Prediction System heating diagrams, terrain alignment provides a powerful way to anticipate daily changes in wind, fuel conditions, smoke movement, and fire behavior.

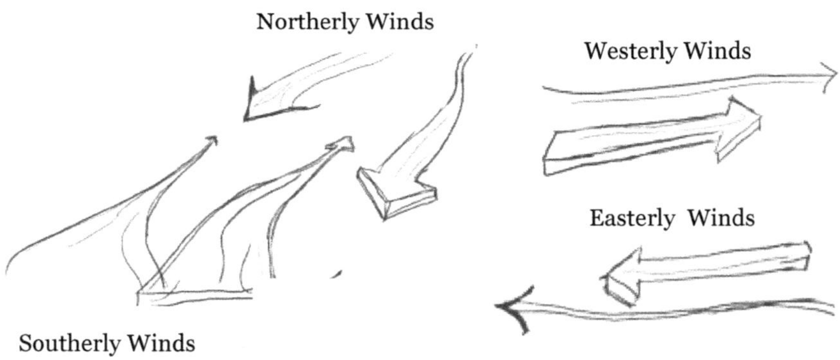

I often use simple wind swirls on maps to indicate airflow direction. These symbols begin with two parallel lines showing where the wind originates and curve outward into a swirl to suggest movement through the landscape. While a simple arrow is often sufficient for indicating direction, three-dimensional arrows can provide a stronger sense of movement through space and help illustrate cloud development, smoke movement, or terrain-driven winds across a landscape. Below are a few examples of simple 3D-style arrows that can be incorporated into maps and landscape sketches.

Building a Visual Map Vocabulary

At the end of the chapter, I share a collection of simple map elements and symbols that can be used to enhance any of the mapping approaches discussed in this chapter. These examples are not intended to be copied exactly. Instead, think of them as visual starting points that can be adapted to fit your own journaling style and the landscapes you observe.

When creating maps in your journal, it is often more useful to sketch recognizable representations of landscape features than to draw them exactly as they appear. A few simple lines may communicate a ridgeline more effectively than a detailed drawing. The goal is not artistic accuracy, but visual clarity.

Effective map symbols allow you to quickly recognize and communicate important features while preserving time and space for observations. There is also value in creating your own map legend or reference page. The process of repeatedly sketching terrain features, vegetation types, clouds, water features, roads, and other symbols helps build familiarity with both the landscape and your preferred methods of recording it. A personal visual vocabulary not only speeds up field sketching, but also creates consistency across your journal, making it easier to compare observations and recognize patterns from one landscape to another.

Building an Internal Landscape Model

Although landform observations often begin with maps and measurements, they ultimately become lived experiences. The angle of a hillside, the exposure of a slope to morning sunlight, or the subtle pooling of cool air in a valley bottom can be felt as much as calculated. Over time, repeated observation allows these features to become part of an internal landscape model—one that informs expectations and observations before conditions begin to change.

When journaling, consider how spatial relationships interact with temporal changes. Elevation influences temperature and moisture patterns. Aspect influences solar exposure and daily heating cycles. Slope affects accessibility, fuel arrangement, and fire behavior. These factors rarely operate in isolation. Together, they form an interconnected framework that shapes the character of a place.

Maps provide structure, but field observations provide meaning. By revisiting places, comparing observations, and updating maps over time, you strengthen your ability to recognize patterns and relationships that might otherwise go unnoticed. It can be especially valuable to return to mapped burn units after fire activity has occurred. Observing post-fire patterns, vegetation responses, or unexpected outcomes often reveals relationships that were not obvious beforehand and helps refine both scientific understanding and intuitive judgment.

Over time, you may find that your maps become simpler rather than more complex. A few contour lines, directional arrows, vegetation symbols, or observation points can communicate far more than a highly detailed drawing. This efficiency reflects not a reduction in observation, but a deeper understanding of place.

Ultimately, the goal of journaling terrain and creating maps is to cultivate a responsive awareness of place. By repeatedly observing how landform interacts with weather, fuels, vegetation, and fire, maps become more than a record of where things are located—they become a tool for understanding how landscapes function

Visual Map Elements Reference

Vegetation/Fuel

Grass brushing against your boots. The crackle of cured leaves beneath each step. Resin rising from sun-warmed pine bark. The cool shade beneath a dense canopy giving way to the heat of an open meadow. Vegetation is often our most immediate experience of a place. It shapes what we see, hear, smell, and feel as we move through a landscape, while also influencing how fire spreads, pauses, or changes character across the land.

In this chapter, we explore vegetation/fuels through multiple scales of observation—from broad landscape patterns and vegetation transitions, to the structure and arrangement of plant communities, to the individual leaves, needles, grasses, cones, and woody materials that make up a fuel bed. Through journaling, we learn to recognize not only what grows in a place, but how vegetation is organized, how it reflects environmental conditions, and how it may respond to fire.

Many of these observations begin with experience. A grassland feels different from a forest. A dense shrub field invites different movement and attention than an open woodland. As you observe vegetation patterns, fuel concentrations, and structural relationships, you begin building a deeper understanding of the landscape and the role that fire plays within it. These observations support storytelling, ecological interpretation, situational awareness, and the study of fire effects.

Seasonal fuel observations and long-term vegetation change are discussed in the Pyrosketchology book. Here, the focus is on the event scale of observation—conditions immediately before, during, and shortly after prescribed or cultural fire. The emphasis is less on identifying every species and more on understanding the patterns, structures, and characteristics that influence how fire interacts with a place.

Numerous formal survey protocols exist for assessing fuels prior to a burn, but they are not the emphasis of this text. Instead, this chapter presents observational and journaling approaches that complement those protocols and support real-time interpretation of fuels during Good Fire events. Data from formal fuel assessments may be available in planning documents and can be incorporated into your journal to strengthen context and comparison.

The illustrations throughout this chapter and the book are intended to inspire curiosity or communicate concepts that may be difficult to explain through words alone. They are not intended as models for what your journal pages should look like. Most of my field journal examples consist of quick sketches, simple diagrams, maps, notes, measurements, arrows, symbols, and observations captured in the moment. The journal examples included throughout this chapter better reflect the speed and practicality of field observation. Focus first on noticing, recording, and reflecting. Artwork can always be refined later, but the opportunity to observe a place in a particular moment is often fleeting.

Landscape Scale—Vegetation Patterns

At the landscape scale, vegetation begins to reveal broad patterns that are often difficult to recognize while standing within a burn unit. Forests give way to shrublands. Shrublands transition into grasslands. Dense vegetation may occupy cooler slopes and valley bottoms, while more open communities occur on exposed ridges and south-facing aspects. These larger patterns provide important clues about how landform, climate, water, disturbance, and fire have interacted over time to shape a place.

This scale of observation helps develop a sense of context before focusing on what you notice when moving through the landscape or start taking measurements. It encourages you to look beyond individual plants and instead notice communities, transitions, and relationships. Areas where vegetation changes abruptly are often among the most interesting places to observe. These transition zones may reflect changes in elevation, aspect, soils, moisture, disturbance history, or management objectives. They can also influence how fire moves across a landscape and help key into areas for situational awareness, like patches of dead and dying trees.

Landscape-scale observations can be made from many perspectives. A high point overlooking a burn unit may provide the best opportunity to observe vegetation patterns directly. In other situations, aerial imagery, topographic maps, Google Earth, drone imagery, fire cameras, or other elevated viewpoints can help reveal relationships that are difficult to recognize from the ground. These tools allow you to step back and see the broader vegetation mosaic before moving into more detailed observations.

As you study the landscape, begin asking simple questions. Where are the largest vegetation communities? Where do vegetation types transition from one form to another? Which areas appear more open or more densely vegetated? Are there patches of dead or stressed trees, snag concentrations, recent disturbances, unusually green areas, or locations where vegetation differs noticeably from the surrounding landscape? These observations often provide valuable context for later discussions of fuel structure, fire behavior, and burn objectives.

Landscape-scale journaling may take many forms. Some observers prefer written descriptions that capture the overall character of the vegetation and its transitions. Others create maps or sketches that emphasize spatial relationships. The Story Maps discussed in the previous chapter provide one useful approach for emphasizing place and experience, while more detailed Study Maps can be used to document vegetation patterns and observation locations. Both approaches help strengthen spatial awareness and provide a framework for observations made later at the scene and close-up scales.

In my examples, I often begin with a simplified burn-unit map, aerial image, or landscape sketch and then emphasize only the vegetation patterns most relevant to my observations. Precision is less important than recognizing relationships. A few broad shapes representing forests, meadows, shrub fields, wetlands, or other vegetation communities are often enough to capture the essential structure of the landscape. Over time, these simplified maps become powerful tools for understanding how vegetation patterns influence the experience of place and the role of fire within it.

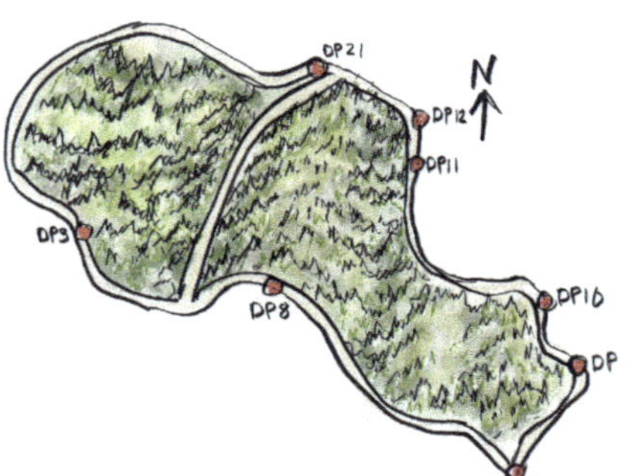

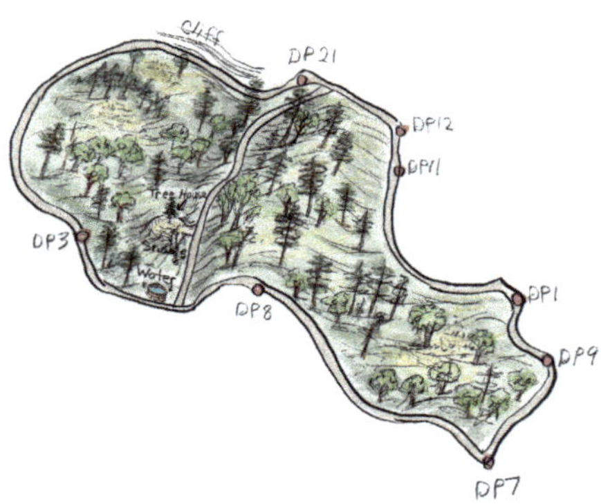

In my examples above, both maps show vegetation patterns within the burn unit. For fuels and vegetation observations, the landscape scale remains focused primarily on the burn unit itself, since our goal is to understand vegetation patterns rather than broader terrain or weather relationships. The map to the left is more representative of actual vegetation cover and provides a stronger sense of overall forested versus open areas. The map to the right accentuates locations where conifer and hardwood trees may be grouped. Both communicate vegetation patterns, but with different intentions. One emphasizes overall vegetation cover, while the other highlights differences in vegetation communities and tree types.

A few tips for sketching vegetation patterns at a scale appropriate for maps and larger landscape observations:

- From an aerial image, elevated viewpoint, or other broad perspective, begin by noticing whether the vegetation can be grouped into general categories such as forests, shrublands, grasslands, wetlands, or barren areas. If forested areas are present, see if you can distinguish between conifer-dominated and hardwood-dominated stands.
- Next, observe how vegetation is arranged across the landscape. Are the trees, shrubs, or grasses sparse and scattered, relatively open, or densely grouped together? Notice where vegetation transitions occur and whether certain vegetation types are concentrated in particular portions of the burn unit.
- When sketching a concentrated vegetation area, such as a forest stand, begin by outlining the overall shape occupied by the vegetation. At this scale, it is often unnecessary to differentiate individual species. However, separating conifer and hardwood forests can provide additional context. I often use pointed shapes and zigzag edges to represent conifer forests, while rounded, cloud-like shapes help convey hardwood forests.

Open Conifer Forest

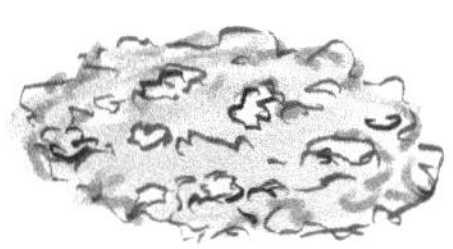

Partially Open
Hardwood Forest

- For more open vegetation communities, you may choose to sketch only a few representative trees or shrubs and leave the surrounding area open. Another option is to lightly indicate vegetation across the entire area while using circles, gaps, or irregular openings to represent meadows, sparse vegetation, or other openings within the larger plant community. See the examples on the next page.

Remember that the goal is not to create a botanically accurate map. Instead, focus on capturing the broad vegetation patterns and transitions that help you understand the character of the landscape and how fire may interact with different portions of the burn unit.

Conifer Forest Hillside

Hardwood Forest Hillside

Vegetation Arrangement & Structure

While landscape-scale observations help reveal broad vegetation patterns and transitions, scene-scale observations focus on the vegetation immediately surrounding you. This is often the scale at which prescribed fire practitioners, observers, and journalers spend most of their time. Rather than viewing a landscape as forests, shrublands, or grasslands, we begin noticing how vegetation is organized, connected, and experienced within a particular place.

At this scale, we observe the structure, arrangement, and character of vegetation. We notice whether shrubs are scattered or continuous, whether grasses form an unbroken layer across the ground, and whether trees create a dense canopy or an open woodland. We begin to recognize how vegetation layers interact and how fuels are connected both horizontally across the landscape and vertically from the surface into shrubs and trees.

The scene scale also provides a bridge between experience and measurement. It is large enough to appreciate the character of a place, yet small enough to document in a journal. Most of the observations, sketches, measurements, and comparisons discussed throughout the remainder of this chapter occur at this scale.

Different people may observe the same vegetation for different reasons. A cultural burner may notice access to food plants, basketry materials, wildlife habitat, or indicators of ecosystem health. A prescribed fire practitioner may notice fuel continuity, vegetation structure, and the potential for fire to move through an area. A wildlife biologist may focus on habitat features, nesting locations, and species diversity. These perspectives are often complementary and can strengthen our understanding of a place.

One challenge of scene-scale observation is deciding how much of the landscape to include. When standing within a burn unit, it is easy to become overwhelmed by the amount of vegetation visible around you. Attempting to document everything often leads to observations that are difficult to compare or revisit. Instead, it is helpful to establish a representative observation area that captures the general vegetation conditions you are interested in studying.

Observation Points

Observation Points (OPs) provide a consistent framework for documenting vegetation structure, fuel conditions, weather, fire behavior, smoke, and fire effects. They help define the scene scale and create a repeatable area for observations before, during, and after a burn.

For journaling purposes, I recommend an observation area of approximately 0.01–0.02 acres. This size is large enough to represent the surrounding vegetation while remaining manageable for sketching, note-taking, and comparison.

To visualize this scale, place a marker such as a water bottle, flagging, or backpack at a central location. Then walk approximately 10–20 feet outward in all directions to establish a circular observation area. The goal is not precision, but consistency. This area becomes the frame through which you observe and document vegetation conditions.

Observation Points help organize both horizontal and vertical observations. Horizontally, they help define representative fuel models, vegetation continuity, concentrations of fuels, and other patterns across the ground. Vertically, they help frame vegetation layers, canopy conditions, fuel size classes, and ecological relationships from the soil surface to the tree canopy. Together, these observations provide a more complete understanding of vegetation as both a living system and a fuel complex.

Observation Points are especially useful when comparing pre-burn and post-burn conditions. By returning to the same observation area, you create a meaningful basis for evaluating how vegetation responds to fire and whether burn objectives were achieved.

Digital mapping tools such as Avenza Maps can help document Observation Point locations and relate them to terrain features, ignition patterns, and burn progression. If photographs are used, consider establishing a consistent photo point by recording the latitude and longitude, camera position, viewing direction, and the primary features intended to be included in the image. Consistency in both observations and photographs strengthens future comparisons and interpretation.

Burn Objectives, Parameters & Desired Change

Once an Observation Point has been established, the next step is understanding why the burn is being conducted. Burn objectives provide context for interpreting vegetation conditions before the burn and evaluating changes afterward. They help transform observations from simple descriptions of what exists into reflections on what may change and why.

Many objectives focus on broad ecological, cultural, or operational outcomes. These objectives help tell the larger story of the burn and provide context for the changes practitioners hope to achieve.

Examples of General Burn Objectives • Improve wildlife habitat

- Maintain or restore fire-adapted ecosystems
- Promote native plant communities
- Support cultural resources and traditional gathering areas
- Reduce hazardous levels of fuels around communities

General objectives are often accompanied by measurable objectives that describe specific vegetation or fuel conditions expected following the burn. Measurable objectives are useful for the study journaling approach.

Examples of Measurable Objectives

- Reduce litter and duff by 50–70% (usually a column measurement including both horizontal and vertical fuel)
- Reduce shrub cover by 30–60%
- Top-kill trees under a six inch diameter (DBH)
- Increase exposed mineral soil by 10%

Burn Parameters (Fire Intensity Goals)

Burn plans also identify desired burn parameters that are expected to produce these outcomes. These parameters often include weather conditions, fuel moisture, fire behavior, and expected fire intensity. For journaling purposes, it can be helpful to think about burn conditions within three broad categories.

Low Intensity –burn parameters that have limited fuel consumption and minimal vegetation change.

Moderate Intensity (often desired conditions)– burn parameters that typically provide sufficient fire behavior to meet most burn objectives, while limiting some more intense fire effects, like higher amounts of tree scorching.

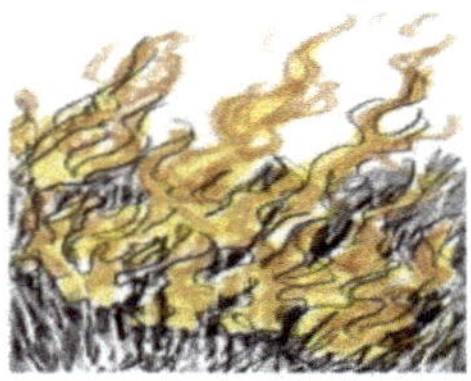

High Intensity – burn parameters with increased fuel consumption, better smoke dispersal, greater vegetation change, and a higher need for strong fire holding personnel and equipment.

Understanding the objectives and desired burn parameters (burn intensity goals) provides valuable context for Story, Study, and Situational Awareness journaling approaches. As you document vegetation conditions, consider not only what exists today, but also what changes may be desired after the burn and what level of fire intensity may be needed to achieve those objectives.

Story

The Story approach helps connect observations to the purpose and meaning of the burn. Consider why this place is being burned, what relationships may be restored or maintained, and how the landscape might look, feel, or function differently after fire. Story observations often emphasize people, place, and ecological relationships.

Study

The Study approach focuses on not only documenting conditions that can be compared before and after the burn, but how different fuel conditions combined with terrain and weather can be used to estimate and observe fire behavior. Consider which vegetation characteristics relate directly to the objectives and what observations may help evaluate success. These observations often include fuel continuity, vegetation structure, canopy conditions, and fuel consumption.

Situational Awareness

The Situational Awareness approach focuses on recognizing when conditions are moving toward or beyond desired burn parameters.

Consider what fire behavior is expected to achieve the objectives and what cues indicate increasing risk.

Horizontal Observations—Fuel Models

Once a representative observation area has been established, the next step is describing the vegetation in a way that helps us understand and communicate potential fire behavior and post fire effects. One of the most common approaches used in prescribed fire and wildland fire management is the fuel model system.

Fuel models are simplified representations of vegetation and surface fuels. Rather than focusing on individual plant species, they group vegetation according to characteristics that influence how fire spreads and behaves. Factors such as fuel type, fuel-bed depth, fuel continuity, and fuel loading help determine how readily a fuel bed may carry fire and the intensity at which it may burn.

Although fuel models are often presented as distinct categories, most landscapes contain variation and transition areas. A burn unit may include several fuel models, and experienced practitioners often consider combinations or modifications that better represent local conditions. The value of fuel models is not that they perfectly describe vegetation, but that they provide a common framework for discussing horizontal fuel fuel arrangements and anticipating fire behavior.

The **Scott and Burgan Fuel Models** (Scott & Burgan 2005, GTR-153) organize these fuels using a few key traits: how deep the fuel bed is, how fine or coarse the fuels are, how much fuel is present, and how it is arranged. These traits interact with wind, slope, and fuel moisture to shape fire behavior on any given day.

In practice, fuel models form a gradient—from light, sparse fuels that carry low-intensity fire, to dense, continuous fuels capable of producing high-intensity, fast-moving fire.

Grass (GR) – from sparse, short grass (GR1) to dense, tall grass (GR9). Fires tend to move quickly, often driven by wind, with relatively short flame residence.

Grass–Shrub (GS) – grasses mixed with shrubs (<50% shrub cover), from light mixtures (GS1) to heavier, more continuous fuels (GS4). Fire behavior begins to build in intensity and duration.

Shrub (SH) – shrub-dominated fuels (>50% shrubs), from light loading (SH1) to dense, heavy stands (SH9). These fuels can produce high intensity fire with longer flame lengths and greater heat release.

Timber Understory (TU) – a mix of litter, grasses, and shrubs beneath a forest canopy, from light (TU1) to dense and continuous (TU5). Fire behavior varies widely depending on understory development.

Timber Litter (TL) – primarily leaf or needle litter, from shallow and patchy (TL1) to deep and continuous with added woody material (TL9). Fire typically spreads along the surface with variable intensity.

Slash–Blowdown (SB) – downed woody material from logging or storms, from light (SB1) to very heavy, tangled fuels (SB4). These fuels can produce high intensity fire and are often difficult to move through and control.

Describing Horizontal Fuel Concentration & Continuity
When a burn plan doesn't specify a fuel model, you can build a working estimate from what you see and feel on the ground. Start simple. Look down. Walk through it.

Pay attention to how horizontally continuous the fuels are, how deep they feel underfoot, and how much effort it takes to move across them. You can describe fuel loading in broad terms:

- **Light** — fuels are sparse and discontinuous; patches of bare soil are easy to see
- **Moderate** — fuels cover most of the ground; most steps land on vegetation or litter
- **Heavy** — fuels are dense and continuous, often including larger material; movement requires stepping over or through logs and concentrated fuels

These observations don't replace formal fuel modeling, but they provide a practical way to connect what you see on the ground with how fire is likely to behave. Over time, this process becomes less about assigning a fuel model number and more about recognizing patterns—how fuels are arranged, how they carry fire, and how they change across the landscape.

Fuel concentration and continuity also influence the conditions experienced by plants, insects, and other organisms. Dense fuels often create different light, temperature, and moisture conditions than open fuels. These small-scale environmental differences, often referred to as microclimates, can provide additional clues about fuel conditions and vegetation structure.

Story Approach to Horizontal Fuel Arrangements

If you are using a Story approach, simple sketches can serve as symbolic representations of a fuel model while helping establish the setting and

Grass

Grass-Shrub

Shrub

Timber-Understory

Timber Litter

Slash/Blowdown

context for what occurred during the burn. Rather than documenting every detail, focus on capturing the character of the vegetation and how it influenced your experience of the place. In this sense, the fuel model becomes one of the characters in the story. See my examples below.

Study Approach to Horizontal Fuel Arrangements

In the Study approach, we want to have data and visuals that help us observe how effective the burn was in meeting objectives. For example, I would add the measurable objectives that are best observed with the horizontal fuel arrangements like; Reduce litter and duff by 50–70% and Reduce shrub cover by 30–60% next to my observation notes and sketches. This way I can compare what I observe with the objectives and fire effects.

I most often use the Study approach in my Good Fire journaling, and one of my most common techniques is a simple horizontal fuel diagram. I create a small square or rectangular frame on my journal page for each of the primary fuel models within the burn unit and use simple symbols to represent different vegetation types, such as trees and shrubs. Light green or yellow shading can be used to indicate grasses or herbaceous vegetation. I may add a few rocks, logs, or other features to better represent what I observed, while keeping the focus on the overall horizontal arrangement of the fuels. See my examples below. The image may look unusual because the upper portion includes a sheet of transparent vellum taped above the diagrams, where I later add fire effects observations and visualizations.

Forest Fuel Model

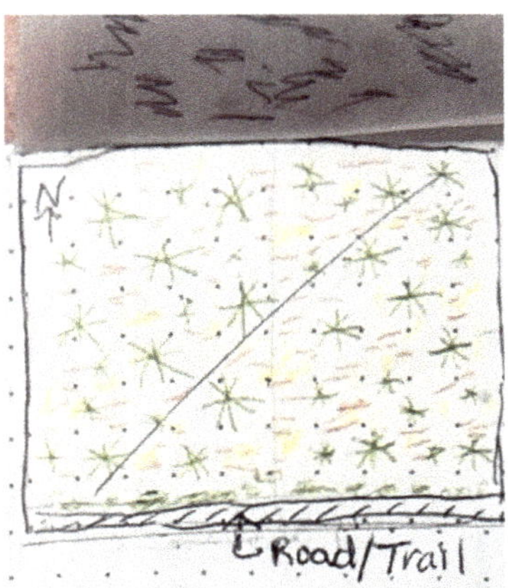

Grass Fuel Model

Situational Awareness Approach

I do not typically create separate Situational Awareness sketches for horizontal fuel arrangements. Instead, I use my burn-unit maps to identify and annotate locations where fuel conditions may require additional attention. Landscape-scale vegetation patterns are especially useful for recognizing areas where changing fuel arrangements may influence fire behavior or risk. Examples include transitions between forest and grassland, dense shrub patches, concentrations of downed woody material, or other areas where fuels become more continuous or capable of supporting higher fire intensity.

Relational Observations

Fuel models provide a useful framework for describing vegetation in ways that relate to fire behavior, but they are not the only way to observe horizontal vegetation arrangements. People often describe and understand the same vegetation through the lens of their experiences, needs, and relationships with the land.

For example, an Indigenous cultural fire practitioner may observe whether vegetation is open enough to move comfortably through, whether food plants are abundant and accessible, or whether basketry and medicinal plants are healthy and of good quality. A dense shrub field may be described not only as a heavy fuel concentration, but also as difficult to travel through or less supportive of desired cultural resources. An open woodland may be valued not only for its fire behavior characteristics, but also for the visibility, access, and ecological conditions it provides.

These relational observations complement fuel models and other measurements by helping us understand how vegetation functions within a living system. As you journal, consider not only how fuels may influence fire behavior, but also how vegetation patterns influence movement, habitat, resource availability, ecological relationships, and human experiences within a place. Such observations often provide valuable context for the ecological indicators discussed in the next section.

Ecological Indicators & Microclimates

Throughout this book, I encourage the use of indicators as a way of observing ecological systems. Indicators help us notice relationships that may not be immediately obvious. Rather than measuring every environmental condition directly, we learn to recognize organisms, patterns, and processes that provide clues about what is occurring within a landscape.

Many of these indicator relationships come from years of field observation rather than formal measurement. They reflect patterns commonly recognized by naturalists, gardeners, farmers, ecologists, trackers, and Indigenous knowledge holders who learn to read environmental conditions through the plants and animals present in a place. Similar approaches can be found throughout natural history, ecological monitoring, traditional ecological knowledge, and place-based observation practices (Leopold, 1949; Kimmerer, 2013). As with all indicators, they should be viewed as clues rather than conclusions and interpreted alongside other observations.

Observing these organisms provides another way of sensing and interpreting fuel conditions. They help reveal whether a fuel bed is open or shaded, dry or moist, warming rapidly or remaining cool, and whether organic matter is accumulating or decomposing. These observations should not be viewed as strict rules, but as additional clues that can strengthen awareness and support journaling observations.

Ants are often associated with open ground, exposed soil, and areas receiving abundant sunlight. Ant mounds and nest entrances can help identify locations where solar heating reaches the soil surface and where vegetation cover may be sparse or discontinuous. Although exposed soil is sometimes viewed negatively, ant activity can indicate healthy ecological processes, including soil mixing, aeration, nutrient cycling, and seed dispersal. Ants are increasingly recognized as useful indicators of ecosystem condition because they respond quickly to environmental change while influencing many ecological processes.

Bees are useful indicators of flowering plants, seasonal warming, and pollinator activity. Areas with abundant bee activity often coincide with vegetation openings that receive sufficient sunlight to support flowering species. Observing bee activity can help identify productive meadow openings, fuel breaks, and other locations where vegetation structure differs from surrounding fuels. In many ecosystems, bee activity also reflects the accumulation of seasonal heat and plant development associated with growing degree days.

Butterflies are excellent indicators of sunlight, warmth, and air movement. Many species depend on solar heating to become active and are frequently observed basking in open areas exposed to direct sunlight. Their presence often highlights warm microsites, vegetation openings, and locations where thermal currents and airflow are concentrated. Butterflies can be particularly useful for noticing transitions between open and sheltered habitats where microclimate conditions change.

Moths are commonly associated with shrub layers, forest edges, and shaded habitats where temperature and moisture conditions differ from nearby openings. Although many species are active at night and often overlooked during daytime observations, moths can serve as indicators of vegetation complexity and the transition zones between open and closed-canopy conditions.

Beetles are frequently associated with woody fuels, bark, decaying logs, litter layers, and standing dead trees. Their presence can draw attention to concentrations of woody material, decomposition processes, and habitat features that may not be immediately apparent when focusing only on live vegetation. Beetles play important roles in nutrient cycling and wood decomposition, helping reveal how fuels are changing over time.

Earthworms are often associated with cooler, moister soils and areas where litter and organic matter accumulate. Their presence can indicate active decomposition, nutrient cycling, and a living soil system beneath the fuel bed. Earthworms are widely used as indicators of soil health because they respond to moisture, organic matter, and soil conditions while contributing to soil structure and nutrient availability (USDA NRCS, 2001). In fire ecology discussions, earthworms are occasionally referenced as indicators of litter consumption, soil disturbance, and post-fire soil recovery.

Taken together, these indicators help expand fuel observations beyond vegetation categories and fuel models. They encourage us to view fuels as part of a larger living system shaped by sunlight, moisture, decomposition, pollination, habitat, and ecological relationships.

Fuel Model Dance & Music Exercise

Most fuel observations focus on what we can see. This exercise invites you to experience fuels through movement and sound. By walking through a fuel model and paying attention to your footsteps, pauses, and pathways, you begin to notice aspects of fuel continuity, arrangement, and structure that are difficult to capture through measurements alone.

Movement and sound can reveal how fuels occupy space. Dense fuels may slow your pace, while open fuels allow more fluid movement. Fine grasses may create soft swishing sounds, while cured stems crackle beneath your feet. Shrubs may require weaving and stepping over obstacles, while forest litter creates its own rhythm of crunches, rustles, and muffled footfalls. These observations help develop a deeper, embodied understanding of fuels and the environments they create.

Although sound can also provide clues about fuel moisture, this exercise focuses primarily on fuel continuity, fuel size, and fuel arrangement. Consider only the sounds produced by your interaction with fuels. Ignore sounds associated with bare soil, rocks, roads, or other non-fuel surfaces. See my fuel model dance and music journal example on the next page and follow the exercise instructions on the following pages. In the next few pages I have a table with word and symbols you might consider using.

Exercise Steps

1. Select a representative area within a single fuel model. Focus on the fuels you are interested in observing rather than attempting to represent the entire burn unit.
2. Establish a straight-line transect through the fuel model. Whenever possible, walk through fuels rather than around them. Step on or over fuels when practical rather than avoiding them. A transect of approximately 50 feet or 25–50 steps is usually sufficient while remaining manageable to document in a journal.
3. Before walking the transect, create a simple cross-section sketch of the vegetation and fuels along the line. This does not need to be detailed. Focus on representing the general arrangement, height, and concentration of fuels. See my example on the previous page.
4. Below the sketch, draw a light and or dashed horizontal line representing your desired walking transect.
5. As you slowly walk the line, pause every few steps to record a small dot (small circle) along the transect line for each footstep.

6. Use a solid/color dot for footsteps that touch fuels and an empty dot for steps on soil or rock.
7. As you move forward, make the dashed/light pencil transect line into a solid line, add any line swerves around large shrubs and trees that you can't step over.
8. Use a half-circle adjoining line to indicate movements stepping over shrubs and symbols for stepping on and over logs.
9. Add short descriptive words, syllables, or sound notes beneath each footstep dot. See the sound syllable, words and visuals on the next page.
10. Include other sounds that arise from your interaction with the fuels, such as stems brushing against clothing, branches tapping together, or seed heads rattling in the wind.
11. After completing the transect, add any additional sketches, symbols, arrows, or visual elements that help communicate the rhythm and flow of the experience.

Footstep & Transect Symbols

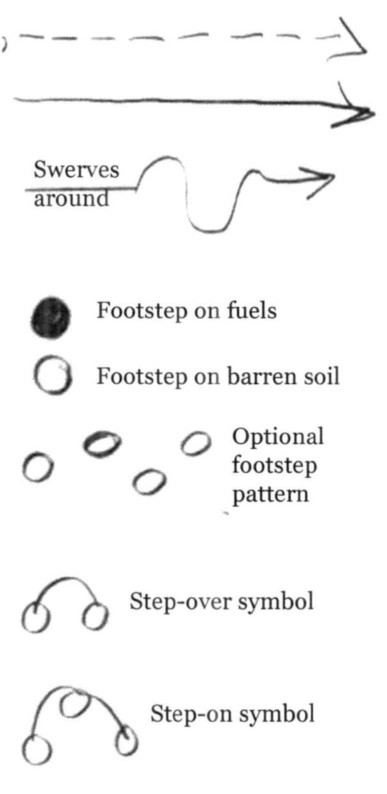

Exercise Reflection

Review your sketch, movement pattern, and sound notes.How did the fuel model influence your movement? Did it encourage a slow, careful pace or a quick and flowing rhythm? Were there pauses, obstacles, or repeated patterns?

Imagine the movement as a form of dance. What type of dance best represents the energy, pace, and flow of the fuels? Is it smooth, energetic, tangled, deliberate, playful, or unpredictable?

Now imagine the sounds as music. What instrument, rhythm, or style of music best reflects the sounds created by your movement through the fuels?

Consider adding any of the ecological indicators discussed in the previous section to provide additional clues about the microclimate and fuel conditions. Are ants, bees, butterflies, worms, moths, or beetles present within your transect? What might they reveal about sunlight, shade, airflow, moisture, decomposition, or vegetation structure? As you reflect on the movement and sounds of the fuels, consider whether these indicators also become part of the dance and music.

Finally, imagine fire becoming the dance partner. How would the movement change? Where would fire move easily? Where would it hesitate, slow, intensify, or change direction? How would the rhythm, pace, and energy of the music change?

Record these reflections in your journal. The goal is not to create a perfect analysis, but to develop a deeper sensory understanding of how fuels influence movement, sound, and ultimately fire behavior.

Footstep & Fuel Sound Words, Symbols & Visuals

Descriptive Word	Sound Syllables	Phonetic Words	Musical Notes
Flowing/Soft	sh, sw, wh, hoo, oo	Swish, whoosh	𝅝 pp ♪
Squeaky/Fabric	ee, eek, sk, ep	Creeek, skereek	~
Dry/Paper	ff, fr, fl, sh, ss	Shfft , frrsh, rustle	♪
Scratchy/Wood	scr, sk, kr, tech, ts	Scratch, krak, snap	♪ >
Crunchy/Brittle	cr, chr, rk, crk , ch	crrrk, chrunch, crack	♪ ♪
Percussive/Strong	th, tok, cl, kl,	clack, pop, bang	♪ > !
Deep/Hollow/ Dull	uh, oh, mm, th	thump, thud, clunk	*f*

Short/Light	♪
Repeated rhythm	♫
Accent/Impact	> ! ~
Sustained/ Flowing	𝅝
Long/Held	𝄢 𝅗𝅥
Low pitch	*f f*
High pitch	𝄞

Sound Visualization & Phonetic Words

swiishhh

ffrrrshhh

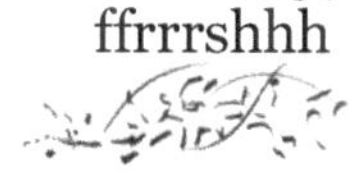

scritch, KRACK, snap

crunch, crackle, crrk

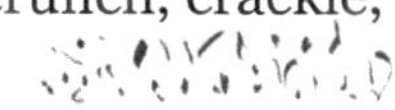

THUMP, crack, BOOM

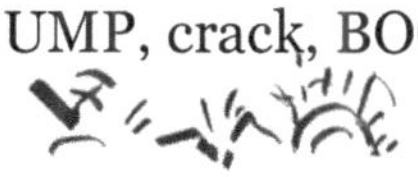

Vertical Fuel Arrangements

The same scene can be observed from multiple perspectives. After examining the horizontal arrangement of fuel models within the observation area, consider how vegetation and fuels are arranged vertically from the canopy to the soil surface. A simple cross-section sketch, similar to the slope-profile sketches introduced in the Landform chapter, provides an effective way to visualize and record these relationships.

Viewing the scene as a vertical fuel profile highlights connections that may be overlooked in plan-view sketches and maps. Tree canopies, ladder fuels, shrubs, grasses, litter, logs, and soil layers each occupy different positions within the fuel profile and may respond differently to fire. Recording observations by vertical layer can reveal patterns in vegetation structure, fuel continuity, habitat conditions, and potential fire behavior.

You can also use the three journaling approaches in slightly different ways:

- **Story** approaches may focus on how people, wildlife, and cultural resources interact with different vegetation layers and how that relates to goals, objectives, and fire effects.
- **Study** approaches compare the height, depth, and arrangement of fuels within the vertical fuel profile. Observations such as tree height, crown base height, ladder fuels, shrub height, litter depth, and duff depth provide a framework for studying how fire moves through different vegetation layers. These observations can also be compared with measurable burn objectives. For example, a burn objective may be to top-kill 40% of small trees while limiting mortality of mature trees to less than 10%, or to reduce shrub heights and ladder fuels while minimizing crown scorch in overstory trees. Comparing pre-fire fuel conditions with post-fire indicators such as scorch height, char height, crown scorch, crown consumption, top-kill, and duff consumption can help interpret fire behavior and estimate whether desired burn objectives were achieved. Vertical observations are often paired with horizontal observations to understand both the behavior and distribution of fire across the burn unit. See the example journal visuals on the previous page.
- **Situational Awareness** approaches may note fuel arrangements that could contribute to increased flame lengths, fire intensity, or hazardous conditions. Ladder fuels, dense vegetation layers, standing dead trees, and concentrations of dead and down material are

examples of features that may warrant additional attention during burn operations.

Vertical Cross Section Diagrams

The two examples above are the cross-section diagram for a transect in the same area as the horizontal diagrams in the previous horizontal diagram section.

The vertical fuel profile provides a useful framework for organizing notes and sketches before, during, and after a burn. The same observation point and scene can be revisited over time to compare changes in fuel structure, fire effects, and burn objectives.

Tree Sketches for Vertical Diagrams

For most vertical fuel profile observations, you do not need to identify every tree species. However, it is helpful to distinguish between conifers (needle-bearing trees) and hardwoods (broadleaf trees). These groups often differ in crown shape, branch arrangement, canopy height, ladder fuels, and the composition of surface fuels beneath them.

Simple tree sketches can help visualize these differences and document how vegetation is arranged within the fuel profile. Conifer stands may contain dense needle litter, deeper duff layers, and different ladder fuel patterns than hardwood stands with leaf litter and grasses. Recognizing these general structural differences can improve observations of fuel conditions, fire behavior, and fire effects.

Shrubs may be represented with simplified rounded or irregular forms. If color is used, darker greens can represent conifers and lighter greens deciduous species. Avoid autumn color palettes, as they may conflict with scorch-related color coding used in fire effects observations.

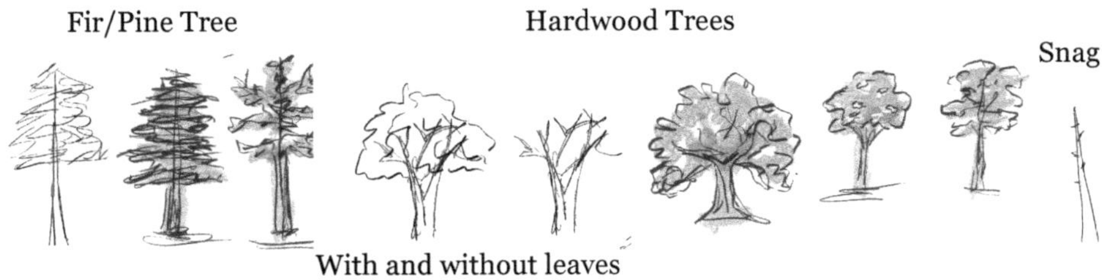

Fuel Measurements

Sketches, maps, and fuel profiles help capture the character and arrangement of fuels within an observation area. Measurements add another layer of information by describing fuel structure, abundance, and size. These observations do not need to be precise inventories. Simple estimates, visual comparisons, and a few field measurements can strengthen comparisons between fuel conditions, fire behavior, fire effects, and burn objectives.

Measurements are particularly useful when revisiting an observation point before, during, and after a burn. Combined with horizontal and vertical fuel observations, they provide a more complete record of how fuels influence fire behavior and how fire alters fuel conditions over time. The visual chart below has the general vertical categories for descriptions, organization, and measurements.

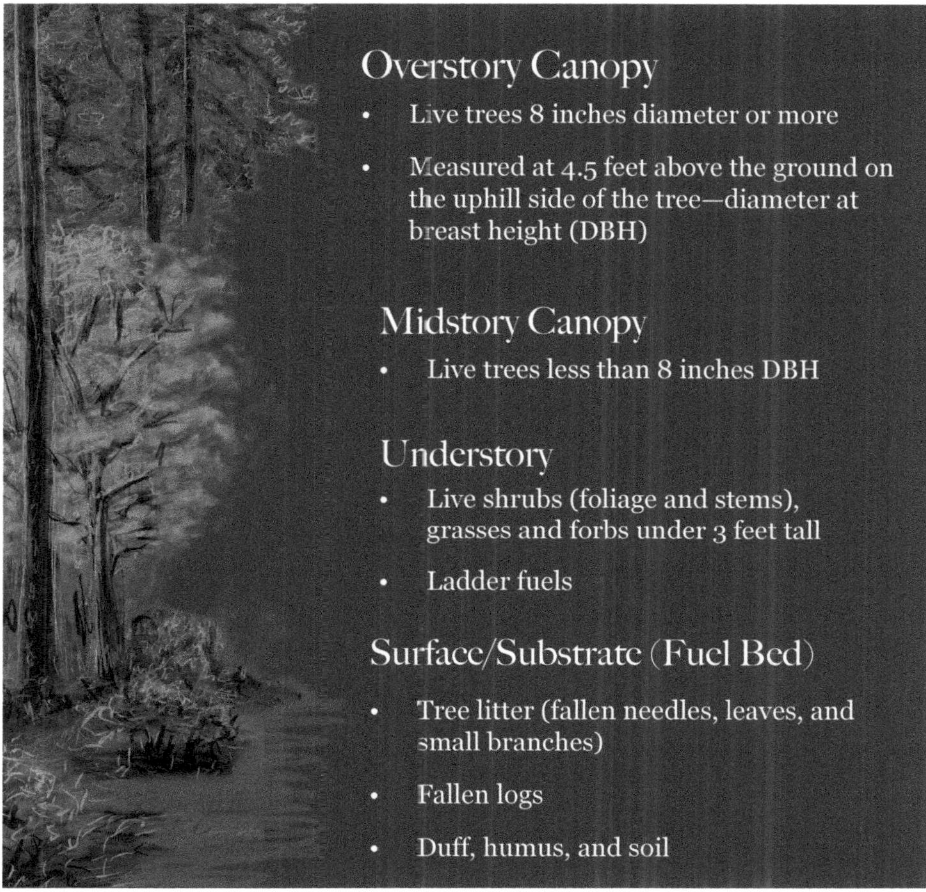

Personal Biometrics

One of the simplest ways to estimate fuel measurements in the field is to use your own body as a reference scale. Personal biometrics transform familiar body dimensions into practical measuring tools that are always available, even when you forget a tape measure.

Before heading into the field, consider measuring and recording a few personal reference dimensions in your journal. Useful measurements may include your overall height, arm span, shoulder width, chest height, hip height, knee height, hand span, fist width, finger width, and the length of your foot. You can add your measurements to the visual on the next page for future reference.

These reference measurements can be used throughout fuel observations. Knee and hip heights can help estimate grass, shrub, and surface fuel heights. Arm spans can help estimate the diameter of larger trees or logs.

Chest height provides a visual reference for diameter-at-breast-height (DBH) measurements. Hand, finger, and fist widths can help estimate the diameter of dead and down fuels. While these estimates are not intended to replace formal inventories, they provide a practical and repeatable method for recording measurements during journaling activities.

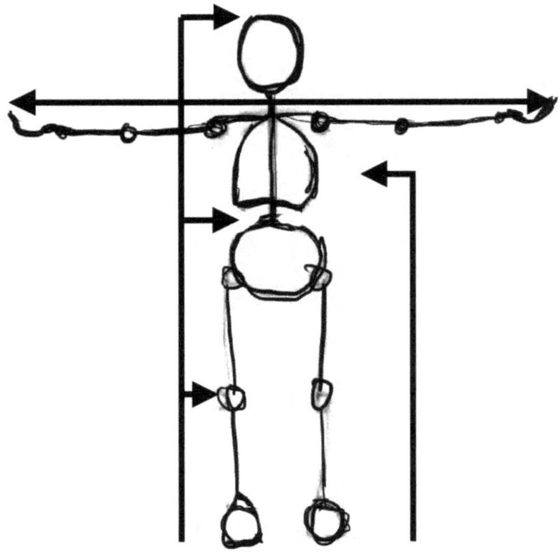

Dead Fuel Diameter Classes

In addition to measuring vegetation structure, dead and down fuels can be grouped into diameter classes. These classes provide a practical framework for observing and describing surface fuel composition. Although they are commonly used in fuel moisture and fire behavior calculations, they can also help organize close-up observations of dead fuels within a journal entry. Their relationship to fuel moisture and fire behavior will be explored further in the Fuels & Weather chapter.

Simple biometric references can help estimate fuel diameters in the field without carrying measuring tools.

1-Hour Fuels – Fine fuels less than ¼ inch in diameter, approximately the width of a pencil lead, matchstick, or small twig. These fuels include grasses, pine needles, leaves, and the finest woody materials found on the surface.

10-Hour Fuels – Woody fuels ¼ to 1 inch in diameter, approximately the width of one or two fingers. These fuels often include small twigs and branches scattered throughout the fuel bed.

100-Hour Fuels – Woody fuels 1 to 3 inches in diameter, approximately the width of a fist or wrist. These fuels commonly include larger branches and small logs.

1000-Hour Fuels – Woody fuels greater than 3 inches in diameter. These fuels include large branches, logs, stumps, and other substantial woody debris that contribute to the overall structure of the fuel bed.

Tree Counts & Diameter Estimates

Tree counts and diameter estimates complement horizontal and vertical fuel observations by adding measurable information about stand structure. These observations can help compare fuel conditions with burn objectives and later evaluate fire behavior and fire effects.

For example, a prescribed burn objective may seek to reduce small-diameter trees and ladder fuels while retaining larger overstory trees. Recording tree counts by size class before and after a burn can help document whether those objectives were achieved.

Diameter at breast height (DBH) is commonly measured at approximately 4.5 feet above the ground on the uphill side of a tree. While direct measurements provide the greatest accuracy, visual estimates are often sufficient for journaling purposes. Personal biometrics, such as hand spans and arm spans, can provide useful reference scales for estimating diameter classes such as 8-inch, 12-inch, or 24-inch trees.

Tree counts, diameter classes, and vertical fuel measurements provide a quantitative complement to sketches and written observations, helping connect fuel structure with fire behavior, fire effects, and burn objectives.

Canopy Cover

Canopy cover is another useful fuel measurement that can be recorded alongside horizontal fuel maps, vertical fuel profiles, and fuel structure observations. Canopy cover represents the percentage of ground shaded by tree crowns, including trunks, branches, and foliage. These observations help explain how vegetation influences sunlight, fuel moisture, wind movement, and fire behavior within an observation area.

Within an Observation Point, estimate canopy cover by looking vertically upward through the canopy. Formal survey methods require multiple samples and statistical averaging, but for journaling purposes one to three representative observations are usually sufficient. See the steps and examples on the next page.

Care should be taken to distinguish between localized canopy shading and broader vegetation patterns. For example, a single observation taken beneath the edge of a tree crown may overestimate representative canopy cover for an otherwise open area. Individual canopy observations remain useful for interpreting micro-site fuel conditions and fire behavior but should not be generalized without additional context.

Estimating Canopy Cover Exercise

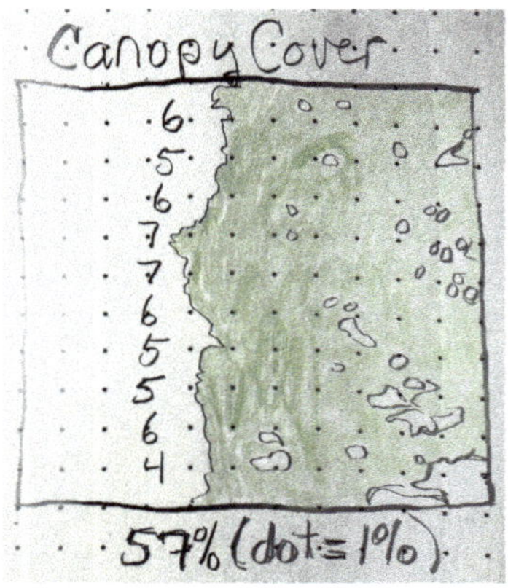

- Use standard-size dot grid paper
- Count eleven dots across and eleven dots down to create a frame.
- This will provide a 10x10 dot grid with 100 dots.
- Each dot covered equals 1% canopy cover.

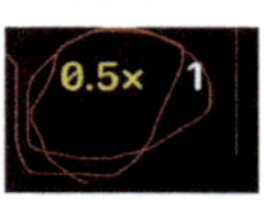

- Use the camera on your smartphone and set the aspect to 1:1 to create a square image that better matches your dot grid diagram.
- Pinch your touch screen to create a wider view, typically 0.5x.
- Take a photo strait up, keeping the camera as level as possible.

- Look at the tree foliage, branches, and trunk in context of how much area of the square it fill.
- On your dot grid, sketch an outline around the edges and openings that represent the canopy cover area.
- Shade in the canopy cover area.
- Count the number of dots covered and total them for your percent canopy cover

Canopy, Cloud Cover, & Fuel Shading

Shading influences ignition probability, fuel temperature, and the moisture content of fine dead fuels. Observations should consider both canopy cover and cloud cover because both reduce solar exposure at the fuel surface. Two primary shading categories are commonly used:

- **Unshaded / Exposed** — Less than 50% combined canopy and cloud cover
- **Shaded** — 50% or greater combined canopy and cloud cover

If canopy cover alone is less than 50%, estimate cloud cover and combine the two values to determine the appropriate category. If canopy cover exceeds 50%, additional cloud cover observations are generally unnecessary because the fuels are already considered shaded.

This combined assessment provides a practical way to evaluate surface fuel exposure during prescribed burns. The exercise that follows demonstrates a simple method for estimating and journaling canopy and cloud cover using a dot-grid journal.

Canopy Cover& Wind Sheltering

Canopy cover also influences wind movement through sheltering effects. As canopy cover increases, wind speeds near the ground are often reduced, affecting flame behavior, smoke movement, and fuel drying.

The following categories provide a useful framework for journaling canopy sheltering effects:

Unsheltered / Exposed Areas — Less than 20% canopy cover

- Typical of grasslands, shrublands, open savannas, and recently disturbed forest areas. Within a 0.01-acre Observation Point, expect approximately 0–2 mature trees.
- Partially Sheltered Areas — 20%–50% canopy cover

Typical of open forest stands with mixed understory vegetation. Within a 0.01-acre Observation Point, expect approximately 2–4 mature trees.

Fully Sheltered Areas — Greater than 50% canopy cover

- Typical of dense conifer or hardwood stands. Within a 0.01-acre Observation Point, expect more than 5 mature trees.

- These sheltering categories provide a simple way to interpret wind-fuel interactions during prescribed burns. Similar canopy cover observations will be used again in the Weather chapter when examining wind behavior, fuel drying, and smoke movement.

Measurements & Journal Format

Throughout this chapter, fuels have been observed from several perspectives. Horizontal observations help describe the distribution and arrangement of fuel models across the observation area. Vertical fuel profiles reveal how vegetation and fuels are organized from the canopy to the soil surface. Fuel measurements add quantitative detail that supports comparisons between fuel conditions, fire behavior, fire effects, and burn objectives. These observations can later be integrated with your maps.

Combining these observations on a single journal page creates a more complete record of fuel conditions than any single observation method alone. Maps, sketches, measurements, and written notes each contribute a different perspective. Together they help document fuel structure, identify patterns, support situational awareness, and create a useful reference for comparing conditions before, during, and after a burn.

Combined Horizontal & Vertical Fuel Observations

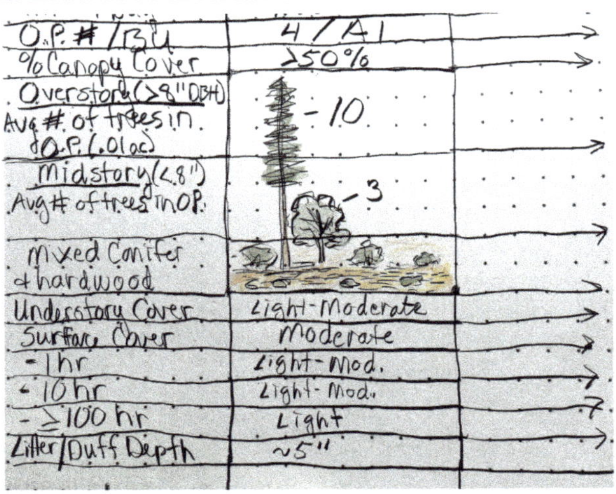

The above journal format provides one example for organizing fuel observations. The column format allows horizontal and vertical observations, fuel measurements, canopy cover estimates, and dead fuel diameter classes to be recorded in a compact and repeatable way. Multiple fuel models or Observation Points can be placed side-by-side for comparison, and the format can be repeated during hourly observations to document changing conditions throughout a burn operation.

Vegetation Through Another Lens

The chart that follows on the next two pages offers another way to observe and think about the vertical fuel profile. Throughout this chapter, vegetation has been examined through fuel models, vegetation patterns, canopy cover, and vertical fuel arrangements. Bird foraging categories provide an additional lens for exploring these same vegetation layers and the resources they contain.

Rather than focusing on birds themselves, this approach encourages closer observation of vegetation structure, composition, and quality. Questions such as "What food resources are present here?", "What plants are producing seeds, fruits, or berries?", "Which vegetation layers support the greatest insect activity?", or "Which plants provide useful materials for people?" can reveal characteristics that may be overlooked when viewing vegetation only as fuel.

This perspective can also help connect vegetation observations to burn goals and objectives. Fuel measurements and vegetation inventories help describe the conditions that influence fire behavior and provide a baseline for evaluating fire effects. Ecological and relational observations add another layer of understanding by encouraging reflection on how vegetation functions within a place and how changes in vegetation structure may influence desired outcomes.

The chart on the next two pages is intended as both an observational guide and a reflection prompt. It expands the vertical fuel profile beyond fuel characteristics alone and encourages consideration of the ecological, cultural, and experiential values associated with different vegetation layers.

Bird Foraging Niche Observations & Relationships

	Bird Foraging Niche	Vegetation Structure Elements
Overstory	**Aerial Insect Feeders** 	• Open airspace for fly-catching • Perches to launch at insects • High vantage points • Edges or openings nearby
Mid Story	**Leaf & Twig Insect Feeders** 	• Leaves and small twigs explored • Higher foliage density for insects • Branching structures to explore and cling • Shrubs in sun or shade
---Understory---	**Shrub Insect & Fruit Feeders** 	• Dense shrubs/branches • Leaves, berries, flowers • Cover and concealment • Connectivity between shrubs • Seasonally changing layers
	Seed Feeders 	• Grasses, sedges, forbs • Seed heads and insects • Ground cover and structure • Seasonally changing layers
Surface	**Ground Insect Feeders** 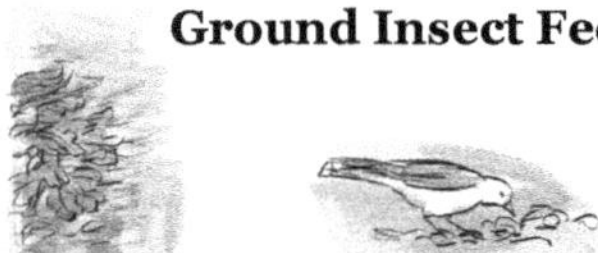	• Leaf litter, needles, duff • Small sticks, bark pieces • Moist, decomposing material • Shaded or protected areas
Subsurface	**Soil Insect/Invertebrate Feeders** 	• Duff layer • Soil invertebrates • Moist soil conditions • Soil texture and structure
Multi-Layers	**Generalist Feeders** 	• Multiple vegetation layers • Diversity of food and cover • Broad, patch, and matrix vegetation conditions

to Fuel Characteristics & Burn Goals/Objectives

Fuel Model Clues	Goal & Objective Clues
• Open or thinned mature forest • Discontinuous crown fuels • Reduced ladder fuels	• Reduced movement of fire into canopy • Better movement for wildlife and people • Moderate surface openings for herbaceous vegetation
• Live crowns with leaves • Moderate density of mid-level trees • Moderate to high live fuel moisture • Forest complexity	• Indicator of healthy leaves and tree vigor • Supports nesting, rearing and diverse species
• Moderate to high shrub density and continuity • Leaves and seeds, fruit, and insects • Shrub diversity and health	• Indicator of healthy and productive shrubs for wildlife and people • Shrub species for basketry • Shrub shelter and cover for wildlife
• Low-moderate overstory canopy • Herbaceous vegetation present • Seasonal growth in resources • Relates to weather parameters	• Seed sources for wildlife and people • Focus for early plant response post burn • Focus for native plant restoration
• Moderate-high litter depth and continuity • Shade and moisture • Decomposing materials	• Indicators for tree litter low-moderate level and openness/access to across, nuts, and fungi • Indicators for soil moisture
• Low-moderate duff depth • Moderate-high duff and soil moisture • Long-term litter accumulation	• Indicator for low-high soil and duff levels and active food web • Indicator for soil moisture and light management for underground foods
• Structural diversity in vegetation • Patchiness and heterogeneity • Multiple fuel models and concentration	• Indicator of diverse vegetation structures and functions • Multiple uses and values to wildlife and people

Weather

The atmosphere is not a passive backdrop to a prescribed burn; it is a living force that shapes the rhythm of every flame and the posture of every smoke column. Weather moves through the landscape with subtle authority, influencing ignition success, moderating fire intensity, and guiding the dispersal of heat and smoke. This chapter invites you to lift your gaze into the vast, mostly invisible currents of the sky and to develop a journaling practice that cultivates both precision and intuition. Through repeated observation, the sky becomes legible—a shifting record of energy, motion, and possibility.

Unlike landforms and vegetation, weather is constantly changing. Conditions observed during the morning briefing may differ substantially from those encountered later in the day. Because prescribed fire operations are guided by burn parameters and weather forecasts, weather observations are not simply descriptions of current conditions. They are opportunities to compare forecasts with reality, recognize emerging patterns, and strengthen situational awareness. A changing sky color, building cloud, shifting wind, or sudden change in humidity may provide clues about conditions developing hours before they appear in weather instruments or operational updates.

The Study and Situational Awareness approaches are particularly important for weather journaling. Forecasts help establish expectations, while repeated observations throughout the burn day help document trends and identify changes. Weather observations become most valuable when they are compared across time. By tracking the sky, clouds, wind, temperature, and humidity, journalers can begin to recognize relationships between atmospheric conditions, fire behavior, and burn operations.

Because this book focuses on journaling during good fire events, the weather elements discussed here are intentionally limited. Burn prescriptions and operational parameters narrow the range of atmospheric conditions that are most relevant to our observations. In *Pyrosketchology*, I explore broader seasonal weather systems, atmospheric processes, and weather observation techniques. Here, our focus remains on the event scale—the hours and days surrounding a prescribed or cultural burn.

Weather measurements should never be taken directly in flames or dense smoke. Instead, measurements should be made in areas representative of where active fire is present or where it is expected to move. Observations are ideally recorded just before the top of the hour so they can be compared with operational updates, forecasts, and changing conditions throughout the burn day.

Burn Parameters

Weather observations during a prescribed burn are most meaningful when viewed through the framework of the burn prescription. Burn parameters establish the acceptable ranges for weather conditions that support safe operations and the desired fire behavior. Rather than simply recording weather measurements, journalers can use these parameters to better understand how changing conditions may influence ignition, fire behavior, smoke, and the likelihood of meeting burn goals and objectives.

The most common weather parameters include air temperature, relative humidity, and wind speed. There are fuel moisture related burn parameters which will be discussed in the next chapter. These measurements are observed throughout the burn day because they continually influence fuel conditions and fire behavior. While the observation I will discuss first in this chapter (cloud cover, sky color, and other atmospheric observations) are not typically included as burn parameters, they can become valuable supporting observations. Over time, these visual cues help develop expectations about changing weather conditions and strengthen situational awareness before instruments indicate significant changes.

General Burn Parameter Weather Criteria & Color Code

Low Intensity	Moderate Intensity	High Intensity
<60°F	60-80° F	>80° F
>40% RH	15-40 % RH	<15%
0-4 MPH	5-8 MPH	>8 MPH

The table on the previous page provides a simple framework for recording common burn parameters. The ranges shown are examples only and should always be replaced with the parameters identified in the burn prescription or operational briefing.

Color coding can help transform measurements into meaningful patterns. Rather than interpreting numbers alone, colors provide quick visual cues that make changing conditions easier to recognize when reviewing forecasts, field observations, and hourly journal entries.

Color Codes for Burn Parameter Tables

Blue — Cooler conditions or lower fire behavior potential
No Highlight — Desired or operational range
Orange — Warmer conditions or increasing fire behavior potential

Using a consistent color scheme throughout your journal helps reveal changing conditions and quickly identify when weather is moving toward or away from the desired operating range.

Forecasts

A weather forecast is more than a collection of numbers—it is a prediction of how conditions are expected to change throughout the day. Reviewing the forecast before arriving at the burn unit establishes expectations for both the Study and Situational Awareness journaling approaches. It provides a starting point for asking, *What observations should I expect today?* and *What changes should I be watching for?*

Creating a simple forecast journal page before a burn encourages anticipatory thinking. Record forecasted temperatures, relative humidity, wind speed and direction, along with notes about cloud cover, and other available weather information for the day before, the day of, and the day after the burn. Color-coding the forecast using the burn parameters can quickly highlight periods when conditions are expected to move toward the upper or lower limits of the prescription.

Include elements such as wind direction arrows, cloud cover, mixing height, or other forecast information available during the morning briefing. As the day unfolds, compare these predictions with your field observations. Notice where forecasts and observations agree, where they differ, and what those differences reveal about local terrain, vegetation, and changing atmospheric conditions.

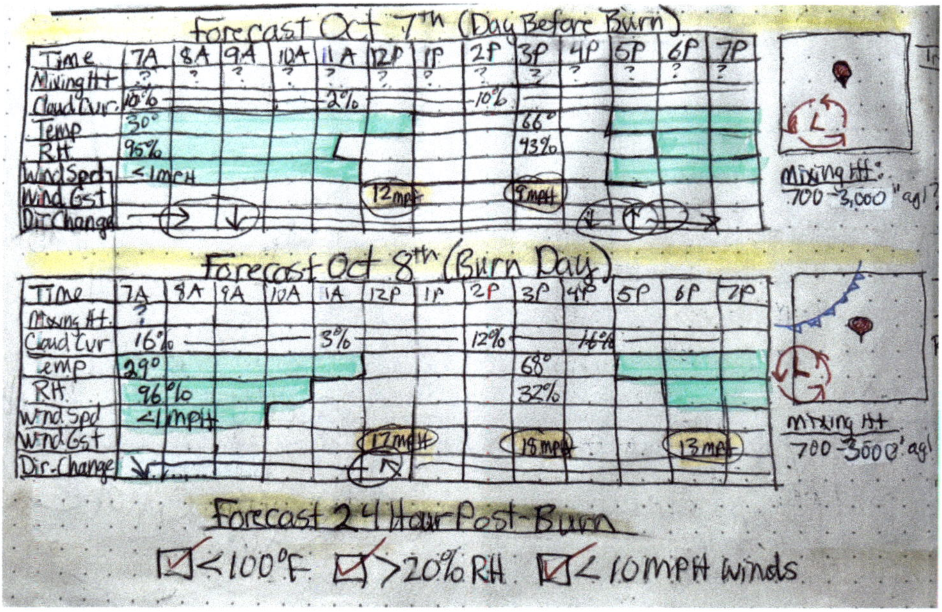

Legend for the abbreviations in my forecast table.

Mixing Ht—smoke mixing height (compare with cloud level).

CloudCvr—cloud cover

Wind Spd—wind speed (average over 1 minute)

Gst—wind gust

DirChange—wind direction change.

Situational Awareness

Situational awareness grows from recognizing patterns rather than recording isolated observations. Throughout the burn day, compare current weather conditions with both the forecast and the burn parameters. Changes in sky color, cloud development, wind, temperature, humidity, or smoke behavior often provide early clues that conditions are changing.

Many of the observations presented throughout this chapter are intended to strengthen this awareness. Repeated observations help develop a personal library of experiences that makes it easier to recognize meaningful weather patterns during future burns. Over time, these comparisons build confidence in anticipating weather changes before they become obvious through measurements alone.

Reading the Sky

The sky is one of the first weather observations made during a prescribed burn and one of the easiest to revisit throughout the day. Before looking at individual clouds or recording weather measurements, pause and observe the overall color and appearance of the sky. Its changing colors often reflect broader atmospheric conditions and provide clues about relative humidity, visibility, atmospheric stability, and approaching weather systems.

Sky color should never be used to predict weather by itself. Instead, it serves as another observational cue that can be compared with the forecast, burn parameters, cloud development, wind, and other weather observations. Over time, these comparisons strengthen both the Study and Situational Awareness approaches, helping you recognize patterns and anticipate changing conditions.

The chart that follows illustrates broad relationships between sky color, relative humidity, sun angle, seasonal influences, cloud development, frontal systems, and atmospheric stability. These relationships are intended as observational prompts rather than precise measurements. The same shade of blue may appear differently depending on the season, elevation, time of day, smoke, haze, or the angle of the sun.

Rather than searching for a single "correct" shade of blue, observe the entire transition from the darkest blue overhead to the lightest colors near the horizon. This gradient often reveals more about atmospheric conditions than any individual color. Notice where the sky becomes lighter, where haze or smoke layers appear, how clouds influence the overall appearance, and whether the sky appears to be changing throughout the day. Repeated observations gradually build a personal understanding of how the atmosphere behaves in the places where you burn.

A Personal Cyanometer

A traditional cyanometer consists of a circular arrangement of blue color samples used to compare the color of the sky. For journaling, I prefer a simple vertical cyanometer strip because it better represents the transition from the zenith to the horizon.

Your cyanometer strip can be drawn, painted, or printed and attached along the edge of a journal page. It only needs to transition gradually from dark blue through medium and light blues to nearly white or cream. The

Sky Color Indicators

Color & RH	Sun Angle & Season	Clouds/ Weather	Atmosphere
• **Overhead /Zenith** • **Humidity less than 30%** • **Dry air**	• Sun high overhead • Deepest blue at midday • Most common in summer and at high elevations	• Cumulus clouds may develop and build • A cold front may have passed the area	• Strong vertical air temperature mixing • Excellent visibility • Generally improving weather weather conditions • Dry/unstable atmospheric conditions
• **Relative Humidity 30-60%** • **Moderate moist air**	• Midday, afternoon • Common in the spring and fall	• Some haze possible • Clouds may be scattered	• Stable, high pressure, neutral atmospheric conditions • Subsidence aloft • Good visibility
• **Relative Humidity 61-85%** • **Moist air**	Common on humid days and late afternoons	• Cloud bases may lower • Clouds may thicken and layer	• Stable atmosphere, • Weaker vertical air temperature mixing • Warm front approaching
• **Near Horizon** • **RH 86-100%** • **Very moist air**	Common in the winter and near sunrise and sunset	• Cloud layers common • Overcast, haze or mist conditions	• Stable atmosphere • Inversion possible with little vertical mixing • Warm front overhead
• **Near Horizon** • **RH 100% Saturated**	Sun angle has less effect when conditions create a more white/gray sky	• Fog, low stratus clouds or marine layers	• Strongly stable atmosphere • Temperature inversion • Poor visibility with smoke trapped

exact shades are less important than having a consistent reference that can be compared with the sky during each observation.

The cyanometer is not intended to estimate exact relative humidity or atmospheric conditions. Instead, it serves as a visual reference that encourages consistent observations and helps reveal relationships between sky color, weather patterns, atmospheric stability, and changing environmental conditions over time.

Imagining the Sky

One of the most valuable ways to use the sky color chart is before stepping outside. Begin by reviewing the weather forecast and asking yourself what you expect to observe.

Will the forecasted relative humidity suggest a darker or lighter blue sky? Is a cold front expected to bring drier air and improving visibility? Will increasing moisture ahead of a warm front produce a lighter, hazier sky? Are stable conditions or an inversion likely to limit vertical air movement? Will cloud development or changing winds alter the appearance of the sky as the day progresses?

After making your field observations, compare what you expected with what you actually observed. Differences between forecasts and observations often reveal the influence of local terrain, elevation, vegetation, smoke, or other microclimate conditions. This simple practice develops anticipatory thinking and strengthens situational awareness throughout the burn day.

Journal Exercise: Sky Colors & Transitions

Step 1: Review the Forecast
Before heading into the field, review the weather forecast and use the sky color chart to imagine what the sky might look like. Record your expectations for relative humidity, cloud cover, wind, atmospheric stability, and any approaching weather systems or frontal passages.

Step 2: Prepare Your Journal Page
Record the location, elevation, date, and time. Include forecast or observed weather information such as relative humidity, wind speed and direction, cloud cover, air quality, or other weather observations relevant to the burn.

Create a simple rectangular frame for your sky sketch. I like to include a small horizon or terrain profile at the bottom of the frame to indicate how much of the sky is obscured by surrounding landforms. This small addition also helps relate cloud observations to the terrain discussed in the Landform chapter.

Step 3: Observe the Sky
Look from directly overhead toward the horizon and compare the observed colors with your cyanometer strip.

Notice:

- The darkest blue overhead.
- The lightest colors near the horizon.
- How quickly the colors transition.
- Haze, smoke, dust, or moisture layers.
- Cloud influences.
- Overall visibility.

Focus on the complete gradient rather than trying to match a single shade of blue.

Step 4: Recreate the Sky Gradient
Using colored pencils, watercolor, markers, or another medium, create a narrow vertical color strip that represents the observed transition from the zenith to the horizon.

Record any noticeable haze layers, smoke, cloud influences, or changes in color intensity. Add a simple horizon silhouette if appropriate.

Step 5: Annotate Your Observations
Beside the sky sketch, record observations such as:

- Relative humidity
- Wind conditions
- Cloud types and cloud cover
- Atmospheric stability
- Forecast information
- Visibility
- Weather systems or frontal influences

Ask yourself:

- Does the atmosphere appear well mixed or stable?
- Is a weather front approaching or departing?
- Is visibility increasing or decreasing?
- How might the angle of the sun influence today's sky colors?

Step 6: Compare and Reflect
Compare your observations with the forecast.

Ask yourself:

- Did the sky appear as expected?
- Which observations matched the forecast?
- What differed?
- How did terrain, smoke, vegetation, or local weather influence what you observed?
- What clues might help you recognize similar conditions during future burns?

Over time, these repeated observations become more than a collection of weather sketches. They create a personal visual reference library that strengthens weather literacy, situational awareness, and the ability to recognize relationships between the atmosphere, fire behavior, seasonal change, and the unique character of each place.

Clouds

Sky color provides a broad impression of the atmosphere, but clouds reveal how that atmosphere is organized and changing. Their height, shape, movement, and coverage offer valuable clues about atmospheric stability, wind patterns, mixing conditions, and approaching weather systems. When observed repeatedly throughout a burn day, clouds become one of the most useful visual indicators for both the Study and Situational Awareness approaches.

Rather than trying to memorize cloud names, begin by asking a few simple questions. At what level of the atmosphere are the clouds? Are they rising, spreading, or flattening? Which direction are they moving? Are they changing throughout the day? These observations often provide more useful information for prescribed fire operations than cloud identification alone. As your observational skills develop, you will naturally begin recognizing the cloud types most commonly associated with changing weather and fire conditions.

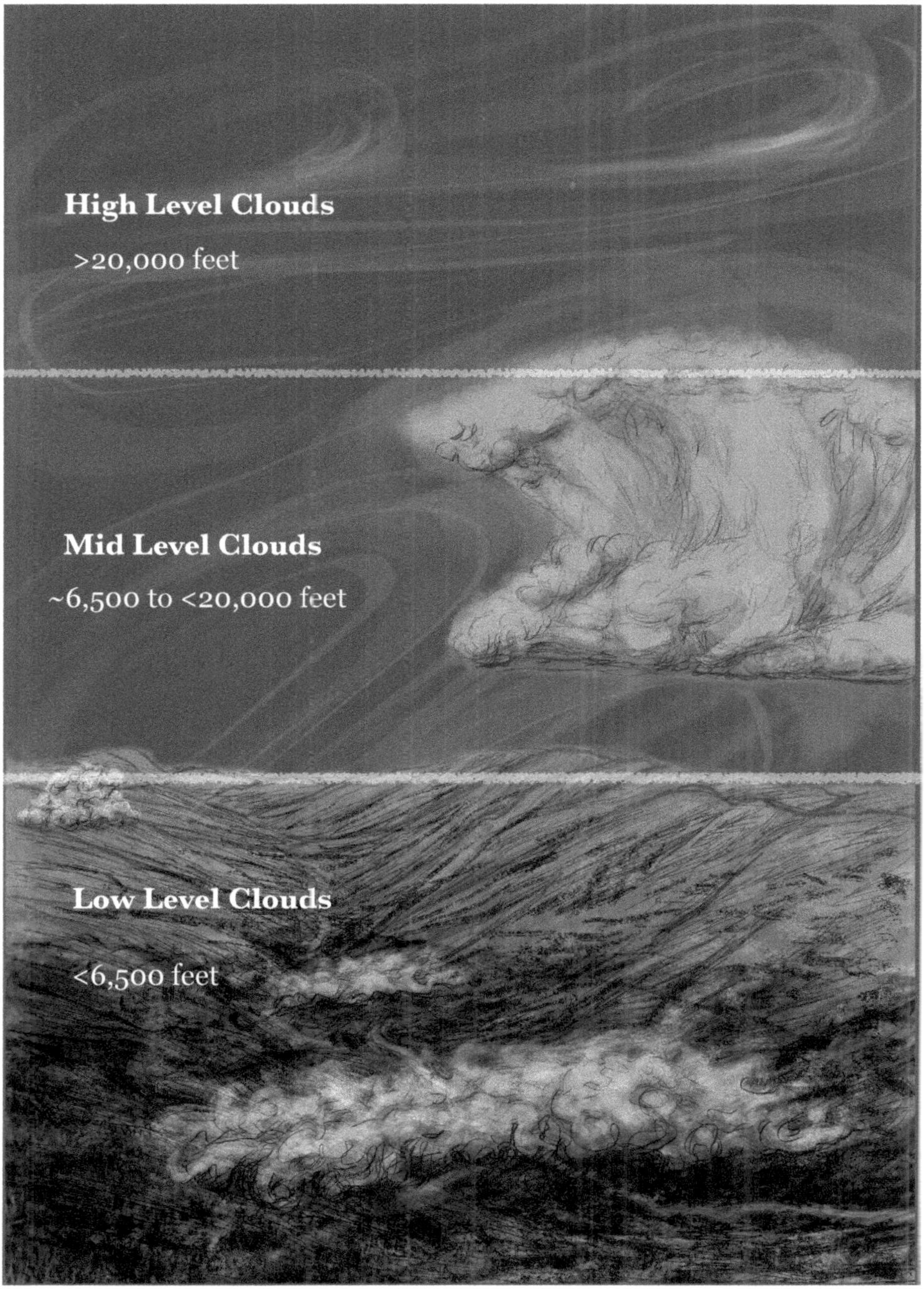

Cloud Levels

We begin by developing a spatial awareness of the sky. Observing cloud levels and movements builds an intuitive understanding of atmospheric structure and vertical air movement. While *Pyrosketchology* explores cloud identification in greater depth, the priority here is understanding how clouds move within the atmosphere and what those movements may suggest about changing weather conditions.

Vertically developing clouds often indicate increasing atmospheric instability as warmer air rises through the atmosphere. This same vertical movement influences smoke dispersion, mixing heights, and, under some conditions, can contribute to greater flame heights and more active fire behavior. Clouds that flatten into layers or slowly settle often accompany more stable atmospheric conditions, where smoke may remain lower and burning conditions become more moderated.

Understanding cloud levels also helps interpret smoke observations and weather forecasts. Weather briefings often include mixing heights, transport winds, or atmospheric stability. Comparing the height of smoke columns with nearby cloud layers provides a practical way to visualize these otherwise abstract forecast values. Recording cloud levels and movements throughout the burn day strengthens situational awareness and builds long-term observational memory.

Hand-to-Eye Cloud Level Estimation

Cloud height can be estimated using a simple hand-to-eye method. Stand facing the horizon and extend your arm fully, keeping one eye open. Align your fingers with distant cloud formations. This technique works best for clouds away from the zenith. Cloud levels shift seasonally, generally rising during warmer months and lowering during colder periods.

- If a single finger width blocks the cloud or cloudlet, it is likely a high-level cloud.
- If two to three finger widths are required, it is likely mid-level.
- If more than three finger widths are needed, it is likely low-level.

This tactile method establishes a physical relationship between your body and the scale of the sky.

High Cloud Level

- Begins around 20,000 feet above ground.
- Ciro-clouds appear thin, string-like, or nearly transparent
- Commercial aircraft cruising altitude often aligns with the lower portion of this level.

Mid Cloud Level

- Occurs between approximately 6,500 and 20,000 feet.
- These alto-clouds can appear hanging along mountain tops or just above. When viewing from higher elevations, these clouds may appear just overhead and low low clouds appearing below like fog.
- They are more opaque and may partially obscure the sun or moon.

Low Cloud Level

- Occurs below approximately 6,500 feet.
- Clouds appear opaque white, gray, or bluish. Cloud types can include nimbus rain clouds and low fluffy early development cumulus clouds.
- Low-level cumulus often have flat bases and darker shading.

Sketching Clouds & Movements

Cloud sketches should be quick, simple, and functional. Begin by capturing the overall character of the cloud rather than small details. Is it building upward, spreading into layers, stretched by strong winds, torn into wisps, or forming wave-like patterns? Add arrows to show movement and include a few notes describing changes since your previous observation.

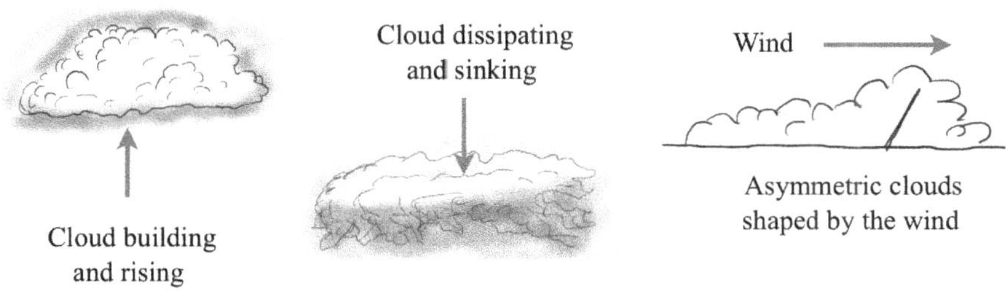

Look for leaning cloud towers, flattened tops, elongated edges, or twisting cirrus that suggest changing winds at different levels of the atmosphere. Simple shading and a few contour lines are often enough to communicate cloud structure. Three-dimensional arrows can also help visualize vertical air movement, wind shear, and the movement of clouds through different atmospheric layers.

The examples below use simple pencil sketches to demonstrate common cloud forms and movements. These are not intended as finished drawings but as visual references for recognizing and recording the cloud patterns most important for prescribed fire observations.

The goal is not artistic accuracy but creating a visual record that can be compared throughout the burn day and across future prescribed burns.

Hourly Cloud Tracker

During active burning, there is often little time for detailed cloud sketches. Instead, I simplify cloud observations into a small set of symbols that can be quickly repeated within my hourly weather tracker.

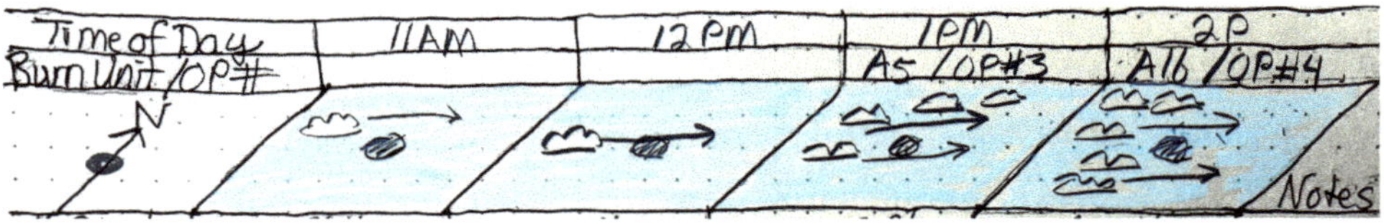

The example above shows the cloud observation row from my hourly tracker. A black dot represents the burn unit, while simple cloud symbols and arrows show the direction and movement of clouds relative to that location. Notice the flat-bottomed, rounded symbol used for cumulus clouds. Additional cloud symbols can be added to represent increasing cloud cover, while layered, wispy, or wave-like symbols can represent stratus, cirrus, lenticular, or other cloud patterns.

Use whatever symbols make the most sense to you, but keep them simple and consistent. Over time, these repeated symbols become a personal visual language that allows you to recognize changing cloud patterns at a glance while leaving room elsewhere on the page for notes or more detailed sketches of significant observations.

I use standard dot-grid paper to quickly organize my hourly trackers and weather tables. The regular spacing makes it easy to maintain consistent hourly columns while keeping cloud observations, weather measurements, and notes aligned throughout the burn day.

Cloud Patterns & Movements

Clouds are constantly changing. Their shape, movement, and development provide valuable clues about atmospheric conditions influencing prescribed fire operations. Rather than identifying every cloud type, observe how clouds change over time. Are they building upward, flattening into layers, becoming stretched by strong winds, or breaking apart? Are they moving toward the burn unit or away from it? Are clouds at different heights moving in different directions? Repeated observations, combined with the weather forecast and burn parameters, strengthen both the Study and Situational Awareness approaches by revealing patterns before changes appear in field measurements.

Think of clouds as relationships rather than isolated objects. Observe how the entire sky is changing. Are clouds becoming more numerous or disappearing? Taller or more layered? Is one portion of the sky becoming more active than another? Vertical growth often reflects rising air and increasing atmospheric instability, while layered clouds generally indicate more stable conditions. Horizontal movement reveals how weather systems and winds are moving across the landscape. Differences in cloud movement between upper and lower layers can provide early clues to changing atmospheric conditions.

Clouds also connect the atmosphere to the landscape. Cumulus clouds commonly develop over warming ridges and mountains, while lenticular and wave clouds often form where strong winds interact with mountain ranges. Looking back and forth between the landscape and sky reinforces the landform observations from the previous chapter and builds a three-dimensional understanding of the atmosphere above the burn unit.

Use the two-page cloud reference that follows as a field guide rather than a cloud identification key. Focus first on what you notice, then consider what those observations suggest about the atmosphere and whether they warrant additional situational awareness. Compare cloud development with the forecast to recognize patterns associated with approaching or departing weather systems.

Repeat cloud observations throughout the burn day. Compare cloud direction, height, movement, and development with forecasted winds, atmospheric stability, and mixing heights. Over time, these observations become part of your personal weather reference library, strengthening your ability to anticipate changing conditions before they become obvious at ground level.

Cloud Observations & Clues for Situational Awareness

Cloud Type (Level)	Suggests (What It May Mean)
Cirrus (High) 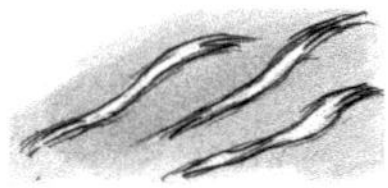 Thin, wispy streaks	• Moisture increasing aloft • Often first sign of an approaching warm front • Strong indicator of weather changes
Cumulonimbus (High) Tall, anvil top Dark tower Precipitation /verga possible 	• Thunderstorm present • Deep atmospheric instability • Moisture in clouds
Altocumulus Castellanus (Mid)  Puffy rows of cloud towers	• Atmospheric instability increasing • Potential thunderstorm development • Moisture and lift building
Altocumulus Lenticularis (Mid) Fairly stationary, smooth, and lens-shaped 	• Strong high-level winds • Strong foehn downslope winds • Unexpected wind shifts • Incoming storm
Towering Cumulus (Low) Crisp outline, flat bottom, cauliflower tops 	• Rapid atmospheric instability • Strong updraft winds
Fair Weather Cumulus (Low) Crisp outline, flat bottom, fluffy sheep tops 	• Earth surface heating • Normal daytime heating • Some atmospheric instability
Stratocumulus (Low) Low, lumpy layers, rolls or patches. Gray underside 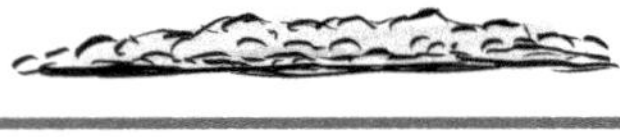	• Stable atmosphere/temperature inversion • Moisture in lower levels • Limited temperature/air mixing

Cloud Observations & Clues for Situational Awareness

Watch For (What to Anticipate)	Sketch/Symbols
• Sky becoming hazier • Additional cloud layers developing • Forecast 12-24 hours out • Potential wind direction changes	
• Lightning potential • Downdraft winds • Sudden wind shifts • Rapid fire behavior changes and potential risk fire behavior	
• Stronger vertical lift and wind updrafts • Afternoon cloud convection • Potential afternoon thunderstorms • Fire behavior intensifying	
• Unexpected wind direction shifts • Strong lee-side winds • Erratic fire behavior • Aggressive fire rates of spread	
• Winds becoming gustier • Faster smoke plume rise • Potential thunderstorm development • Increasing fire behavior	
• Cloud tops may build rapidly • Watch the time clouds are building • Potential for increasing fire behavior	
• Lowered smoke mixing height • Persistent cloudiness • Reduced fire behavior	

Cloud Cover

Cloud cover describes the fraction of the sky obscured by clouds and is commonly estimated in eighths. Although simple, it is one of the most useful weather observations made during a prescribed burn. Changes in cloud cover influence sunlight, fuel shading, air temperature, and the warming of fine fuels throughout the day. Combined with the forecast, repeated cloud cover estimates also strengthen situational awareness by helping you recognize changing atmospheric conditions.

In the Fuels chapter, canopy cover was used to estimate fuel shading beneath vegetation. Cloud cover becomes the atmospheric component of that same observation. Together, canopy and cloud cover influence the amount of solar radiation reaching surface fuels and provide context for later discussions of fuel moisture and ignition probability.

Cloud cover is commonly described using a few simple observational categories. These descriptive terms provide a quick way to communicate changing sky conditions, while the eighths estimate provides a more quantitative measurement for journaling and comparison.

Cloud Cover Categories

 •**Clear Sky** — No clouds visible (0/8)

 •**Few Clouds** — Less than 1/8 to 2/8 cloud cover

 •**Scattered Clouds** — 3/8 to 4/8 cloud cover

 •**Broken Clouds** — 5/8 to less than 8/8 cloud cover

 •**Overcast** — Complete cloud cover (8/8)

Estimating Cloud Cover Exercise

Estimating cloud cover becomes easier with practice. Rather than judging the entire sky at once, divide your observations into four directions: north, east, south, and west.

Using a dot-grid journal, create a wide rectangular page divided into four sections. Begin each section with a simple horizon sketch that represents the surrounding terrain. Including the landscape reminds you how ridges, valleys, forests, or other landforms influence what portion of the sky is visible from your Observation Point. This small addition also continues building the story map of place developed in the Landform chapter by connecting the atmosphere to the landscape below.

Next, sketch the cloud cover observed while facing each direction using simple cloud symbols. After completing all four views, mentally combine them into a single view of the sky and estimate the fraction of cloud cover in eighths.

The exercise is not intended to produce perfect measurements but to develop a consistent method for observing and comparing cloud cover over time. Like many Pyrosketchology exercises, repetition is more important than precision. Repeated observations strengthen your visual estimates while building connections between the forecast, cloud development, fuel shading, and changing atmospheric conditions throughout the burn day.

Date Cloud Cover Location Time

In this example I am estimating about 3/8ths of the sky has cloud cover or scattered clouds.

Wind

Unlike clouds and sky color, wind cannot be observed directly. Instead, we experience it through its effects on the landscape and ourselves. We see grasses bend, leaves flutter, and tree crowns sway. We hear branches creak, dry vegetation rattle, and the changing voices of the wind moving through different fuel types. We feel changing temperatures, moisture, and pressure against our skin. These visual, auditory, and tactile observations often reveal as much as a wind measurement taken with an instrument.

For prescribed fire practitioners, wind is one of the most influential weather elements because it affects ignition patterns, flame length, rate of spread, smoke movement, and operational tactics. Yet wind is rarely constant. It changes with terrain, vegetation, time of day, and approaching weather systems. Learning to recognize these changes through repeated observation strengthens both the Study and Situational Awareness approaches while developing a deeper understanding of how weather interacts with fuels and topography.

This section begins with burn parameters and forecasted winds before moving into ways of observing wind through vegetation, sound, and movement across the landscape. Like the other chapters in this book, the goal is not simply to collect measurements but to build a personal reference library that helps you recognize changing conditions during future burns.

Burn Parameters for Wind

Wind observations become most meaningful when viewed through the burn prescription. Burn parameters define the range of wind conditions that support safe operations and the desired fire behavior. Rather than simply recording wind speed, compare each observation with the operational range identified during the morning briefing. Is the wind within the desired range? Is it becoming stronger or weaker? Is the direction changing? These comparisons provide context that raw numbers alone cannot.

Most prescribed burns define acceptable ranges for both wind speed and wind direction. Wind speed influences flame length, rate of spread, spotting potential, and smoke movement, while wind direction determines where fire and smoke are expected to travel. Even when wind speeds remain within prescription, a significant shift in wind direction may change fire behavior and operational tactics.

The general wind categories below provide a simple framework for journaling. Whenever possible, replace them with the specific burn parameters for the prescribed fire you are observing.

- **0–4 MPH** – Low wind speeds, generally associated with lower-intensity fire behavior.
- **5–8 MPH** – Moderate wind speeds that commonly support desired prescribed fire behavior.
- **Greater than 8 MPH** – Higher wind speeds that may increase fire intensity and require additional situational awareness.

Like the other burn parameters in this chapter, I find it helpful to color-code these ranges throughout my journal. Blue represents lower wind speeds, no highlight represents the desired operational range, and orange highlights increasing wind speeds where additional awareness may be needed. These colors create quick visual cues that make changing conditions easier to recognize when reviewing forecasts, hourly observations, and burn operations.

Common Wind Types

Many local wind patterns occur with predictable daily heating and cooling cycles or as weather systems move across the landscape. Understanding these common wind types provides a useful framework for anticipating how wind may behave before arriving at a burn unit. Combined with the weather forecast, terrain maps, and burn parameters, they help establish expectations for both the Study and Situational Awareness approaches.

For example, a forecast calling for light afternoon winds may suggest the development of upslope or up-valley winds as the landscape warms. Likewise, a forecasted cold front may indicate stronger, more sustained winds that differ from the typical daily terrain winds. By comparing the forecast with these common wind patterns, you begin each burn with questions rather than assumptions: *What wind patterns do I expect? Where might they be strongest? How might the terrain modify them?*

The common wind types below are not rigid rules. Local terrain, vegetation, weather systems, and seasonal conditions often modify both wind speed and direction. As you build your own weather journals, compare these generalized patterns with your field observations and note where they agree—or where local conditions create something different.

Wind Types (Typical Progression)

- **Downslope:** 2–5 MPH; evening and nighttime flow downhill as air cools.
- **Upslope:** 3–8 MPH; late morning and afternoon flow uphill as slopes warm.
- **Land Breeze:** 3–10 MPH; nighttime flow from land toward water.
- **Down Valley** : 5–10 MPH; nighttime drainage winds following valley bottoms.
- **Up Valley:** 10–15 MPH; afternoon winds moving up valleys as heating increases.
- **Sea Breeze:** 10–20 MPH; daytime flow inland from large water bodies.
- **Warm Front:** 10-20 MPH in a broad southeasterly flow. Often produces broad, sustained winds with gradually increasing moisture and cloud cover.

- **Cold Front:** 15-30 MPH in a broad northeasterly flow. Often brings stronger, shifting winds with cooler, drier air following frontal passage.

The isometric terrain sketches and modified Campbell Prediction System diagrams introduced in the Landform chapter provide a helpful way to visualize these wind patterns. Comparing forecast winds with these landscape relationships creates a mental picture of how winds may change before you begin making field observations.

Wind Through the Landscape

Forecasts describe winds across a broad area, but the wind experienced within a burn unit is rarely uniform. Terrain, vegetation, and exposure continually modify wind speed, direction, and turbulence as air moves across the landscape.

Ridgetops are typically the most exposed to prevailing winds, while valleys, draws, and dense forests often reduce surface wind speeds. Mid-slopes commonly experience changing wind patterns as daily upslope and downslope flows develop. Saddles, canyon mouths, and mountain gaps can accelerate winds as air is funneled through narrow openings, while sheltered slopes may experience lighter or more variable winds.

These local variations explain why observations made at one location may not represent conditions elsewhere within the same burn unit. As you move through the landscape, compare what you feel with what you see. Notice where vegetation begins moving differently, where smoke changes direction, or where the wind suddenly becomes warmer, cooler, stronger, or calmer. These observations help connect the forecast to the landscape you are experiencing.

Butterflies provide another subtle indicator of changing wind and heating conditions. As sunlight warms the landscape, many species become more active, taking advantage of gentle thermals and sheltered air currents within the vegetation. Their movements can prompt you to notice small-scale air movement and heating patterns that may be overlooked when focusing only on tree tops or wind instruments. Like the bird observations in the Fuels chapter, butterflies offer another ecological lens for observing relationships between weather, vegetation, and place.

Wind Measurements & Adjustments

Field measurements provide an important reference for comparing your observations with forecasts and burn parameters. Wind can be measured using instruments such as a Kestrel or sling psychrometer, or obtained from nearby weather stations and forecast products. Whenever possible, make measurements while facing into the wind in an open area away from immediate obstructions.

Weather forecasts typically report **20-foot winds**, representing conditions above the surrounding vegetation. Winds measured within a burn unit are often **mid-flame winds**, where vegetation, topography, and canopy cover reduce or redirect airflow. Understanding this difference helps explain why observed winds may differ from forecast values.

The wind adjustment chart that follows summarizes the standard factors used to estimate surface winds from forecasted 20-foot winds. Before using the chart, estimate canopy cover and slope using the observation techniques introduced in the previous chapters. These measurements not only improve wind estimates but also reinforce the relationships between vegetation, landforms, and weather that shape fire behavior throughout a prescribed burn.

Wind Through Vegetation

Vegetation is one of the easiest and most reliable ways to observe wind in the field. Long before you reach for a weather instrument, grasses, shrubs, leaves, and tree crowns are already responding to changing wind speeds and directions. By learning to recognize these movements, you can build a personal visual reference that complements wind measurements and strengthens situational awareness during prescribed burns.

The exercises on the next few pages are designed to calibrate your observations rather than your artistic skills. Instead of memorizing a standardized wind scale, you will compare vegetation movements with measured wind speeds and the burn parameters for your prescribed fire. Over time, these repeated observations become a location-specific reference library, helping you recognize when wind conditions are approaching—or moving beyond—the desired operating range.

Wind Speed Adjustment

Based on Slope, Canopy Cover Sheltering, & Fuel Model

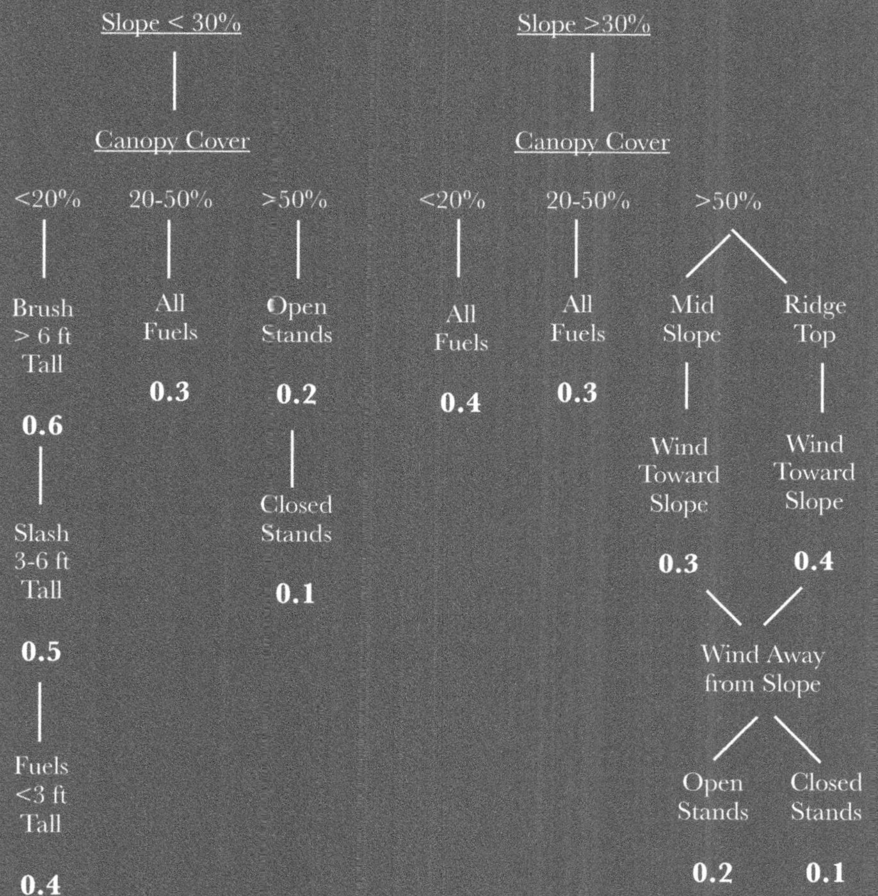

Multiply the open/20-foot wind speed by the appropriate number in the flowchart for an adjusted wind speed factor at eye-level.

<20% Canopy Cover (Unsheltered/Exposed)—Grass, Grass-Shrub, Savanna-like and more open forested areas.

20-50% Canopy Cover (Partially Sheltered)— Fairly open mixed conifer stands, open lodgepole or ponderosa pine stands, or open hardwood stands.

>50% Canopy Cover (Fully Sheltered)— dense short-needle conifer stands such as spruce or fir, and dense hardwood or pine stands.

Forest Edge Exercise

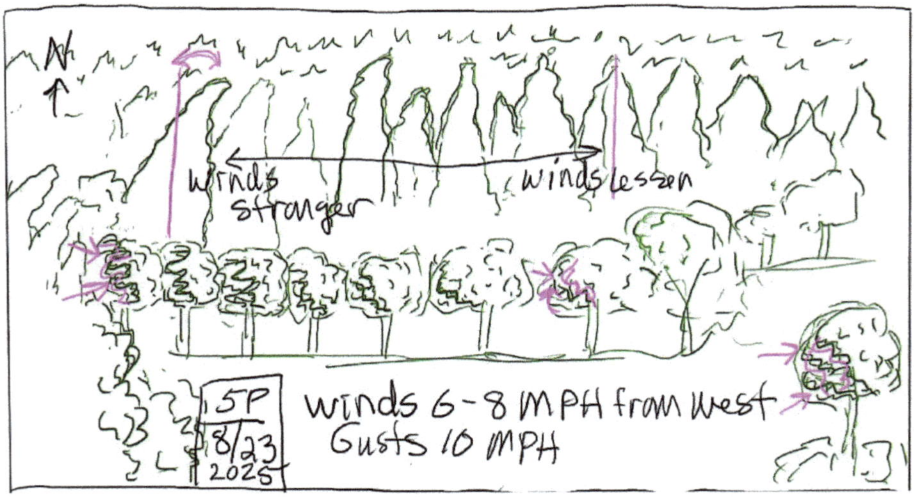

This exercise develops a broad-scale understanding of how wind moves through vegetation across the landscape. It works especially well with the modified burn parameter wind categories discussed earlier in this chapter.

Step 1: Select an Observation Point

Look for an opening where you can clearly view the edge of a forest or other vegetation type. Whenever possible, position yourself so the wind is moving across your view from left to right or right to left. If the wind is directly in your face or at your back, vegetation movement will be more difficult to visualize, but the exercise can still be completed.

Step 2: Record the Context—Before sketching, record:

- Date and time
- Observation Point or location
- Slope or terrain position
- Direction you are facing
- Forecast and/or measured wind speed and direction
- Vegetation type or dominant species (at least note whether it is primarily conifer or broadleaf)

These observations become more meaningful when compared across seasons, fuel models, and burn parameter categories.

Step 3: Sketch the Vegetation—Create a small to medium-sized landscape-oriented frame. Sketch the outline of the tallest canopy trees when they are at their greatest lean from the wind. For distant tree tops, a simple leaning trunk line is often enough. For trees closer to you, sketch the general canopy outline and use a few loose movement lines to indicate trunk and branch movement rather than drawing every branch.

Compare how the amount of movement changes as wind speeds move from the lower, moderate, and upper burn parameter ranges.

Individual Tree Exercise

After observing the landscape, narrow your attention to a single tree growing in a more open or exposed location.

Step 1: Select the Tree
Choose a tree where the wind is moving across your view from left to right or right to left.

Record the tree species if known, or simply note whether it is a conifer or broadleaf tree.

Step 2: Create the Framework
Using a pink, purple, or blue pencil, sketch a large cross with a longer vertical line extending below the center. Then add a large "X" centered on the cross to create a simple directional reference.

Step 3: Sketch the Tree Canopy
Using a green pencil, sketch the basic outline of the tree canopy and the upper portion of the trunk.

Rather than drawing every branch, select one representative branch closest to one of the reference lines. Sketch that branch at the point of its greatest movement during stronger gusts.

Step 4: Estimate Scale of Movement
Hold up a pen in vertical position to the tree trunk. Angle the pen to the left or right to see the degree of angle as the tree and branches move.

Add arrows showing both the direction and amount of branch movement. Compare the observed movement with the measured wind speed and the burn parameter category.

As you repeat this exercise under different weather conditions, you will begin recognizing how different tree species respond to changing wind speeds. Even the practice of just holding up a pen and using the movement of your writs can enhance your observation skills.

Leaf Movement Exercise

The smallest vegetation movements often appear before larger branches begin to sway. This exercise helps train your eye to recognize those subtle changes.

Using the same tree, repeat the reference cross and "X" used in the previous exercise.

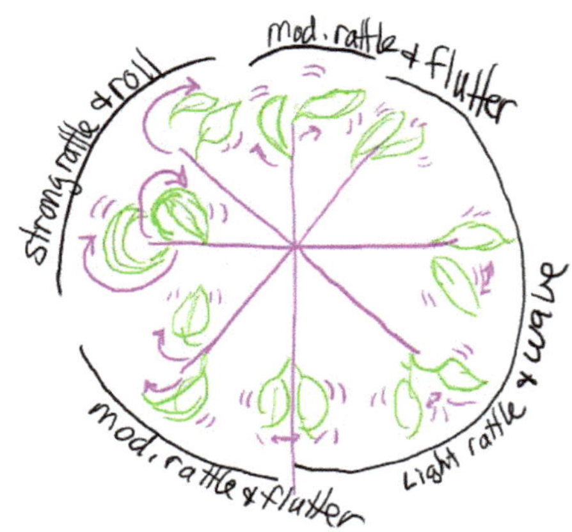

<u>Step 1: Sketch Leaves</u>
Sketch one or more representative leaves or clusters of needles from different parts of the canopy in positions of movement.

<u>Step 2: Observe Movements</u>
Observe how the leaves respond during stronger gusts.

Do they: Lift? Flutter? Wave? Rotate? Flap? Fold?

<u>Step 3: Add Symbols</u>
Add arrows to indicate movement direction and use short or long accent lines to show the amount of shaking or vibration.

Step 4: Add Words
Finish by adding a few descriptive words about both the movement and the sounds you hear. Over time, these observations become another way of recognizing when winds are approaching important burn parameter thresholds.

Wind & Sound

Wind is often heard before its effects become obvious in surrounding vegetation. Learning to listen carefully adds another layer of situational awareness that complements visual observations and weather measurements.

Different fuel types produce distinctive sounds as wind increases. Grass may whisper or hiss. Dry shrubs may scratch or rattle. Branches may creak before tree crowns begin to sway. These sounds vary with fuel type, moisture, season, and wind speed, creating another way of recognizing changing conditions.

During Fire Effects Monitoring training, I noticed that dry mule's-ear plants produced only a faint scratching sound under lighter winds but shifted to a distinct rattling sound as wind speeds approached the upper end of the desired burn parameters. That simple auditory change became an early reminder to pay closer attention because increasing wind often preceded noticeable changes in fire behavior.

When the wind reaches 6 MPH, the mule's ears start a raucous chatter.

Deep Listening Practice

Pause several times throughout the burn—before ignition, during active burning, and after fire passage—and intentionally listen.

Notice:

- Duration of the sound
- Repetition of sound elements
- Volume or intensity
- Frequency or pitch

- Patterns in the sound (steady, rising, falling, pulsing)
- What the sound reminds you of

Expand your listening beyond the wind itself. Notice how terrain shapes sound as it moves across open ground, through forests, into canyons, or along rocky slopes. Compare how different fuel models respond as grasses whisper, shrubs rattle, or timber creaks beneath increasing winds.

Also consider how weather conditions influence what you hear. Temperature, humidity, and changing wind speeds affect how sound carries across the landscape, while birds, insects, machinery, and human activity become part of the changing soundscape throughout the day.

The examples on the following page illustrate several simple ways to visualize these sounds using marks, symbols, words, and musical ideas. The goal is not to create precise recordings, but to build a personal sensory reference that strengthens your awareness of changing wind and fire conditions.

Experiencing the Wind Through Place

Weather is rarely experienced the same way throughout a burn unit. As you walk across ridges, through forests, into openings, or along drainages, wind speed, direction, temperature, humidity, cloud cover, and sound continually change. This final exercise brings together observations from the Landform, Fuels, and Weather chapters into a single experience of place.

Before beginning your walk, review the weather forecast and identify a period when changing wind conditions are expected. As you move through the landscape, notice how the wind feels against different parts of your body. Does it strike one side of your face? Does it strengthen as you leave the forest? Does it become calmer beneath the canopy or accelerate through a mountain gap or saddle?

Use a simple stick figure to represent yourself at different locations along your route. Sketch your hair or clothing to indicate wind direction and add arrows showing both the wind and your direction of travel. Record cloud movement, vegetation movement, sounds, and any noticeable changes in temperature or humidity as you move through different landforms and vegetation.

This exercise is more than a map of changing wind. It becomes a story map of weather and place, connecting forecasts, landforms, clouds, vegetation, and your own sensory experience into a single journal page. Over time, these walks build a richer understanding of how weather behaves across a landscape and strengthen both your situational awareness and your ability to anticipate changing fire conditions.

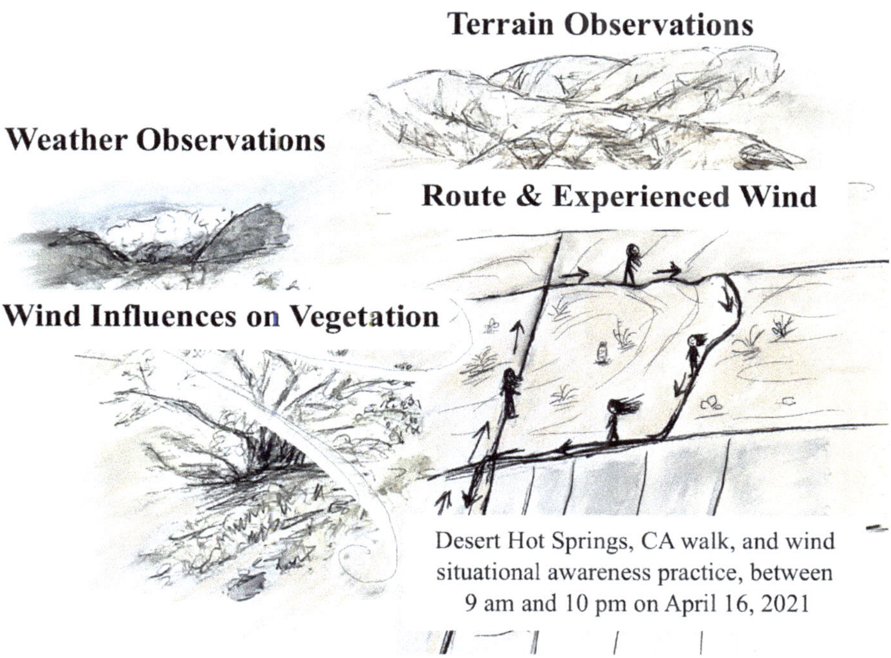

Desert Hot Springs, CA walk, and wind situational awareness practice, between 9 am and 10 pm on April 16, 2021

Forecast Weather During Potential Walk

Time	9:00 am	10:00 am	11:00 am	12:00 pm	1:00 pm	2:00 pm	3:00 pm
Wind Speed & Direction	11 mph westerly	13 mph westerly	17 mph westerly	19 mph westerly	20 mph westerly	22 mph westerly	23 mph westerly
Temperature	68° F	71° F	79° F	76° F	78° F	79° F	79° F
Relative Humidity	45%	38%	33%	30%	28%	27%	27%

Temperature & Relative Humidity

Throughout this chapter, you have learned to observe weather through sky color, clouds, wind, and other sensory cues. Temperature and relative humidity provide the measurements that help explain and verify many of those observations. Together they form the quantitative foundation that complements the visual and experiential observations developed throughout this chapter.

Measure temperature and relative humidity using a field weather instrument or reliable forecast data. Whenever possible, take measurements in the shade or with your back to the sun to minimize the influence of direct solar heating. Like the other weather observations in this chapter, compare each measurement with the prescribed burn parameters rather than viewing the numbers in isolation. Continue using the color-coding system introduced earlier in the chapter to quickly distinguish conditions below, within, and above the desired burn parameters.

Forecast Comparisons

By this point in the chapter, the forecast has already helped frame your observations of the sky, clouds, and wind. Temperature and relative humidity complete that comparison. Record the forecast alongside your field measurements and note where they agree—or where local terrain, vegetation, canopy cover, cloud cover, or other environmental factors create different conditions. These comparisons strengthen both the Study and Situational Awareness approaches while building confidence in interpreting forecasts during future prescribed burns.

Journaling—Hourly Weather Tracker

The hourly weather tracker brings together the observations developed throughout this chapter. Record temperature, relative humidity, wind speed and direction, gusts, cloud movement, cloud cover, and other atmospheric observations within the same hourly format. Continue using your color-coding system to highlight conditions that fall below, within, or above the desired burn parameters, providing an early visual cue for changing fire behavior or operational conditions.

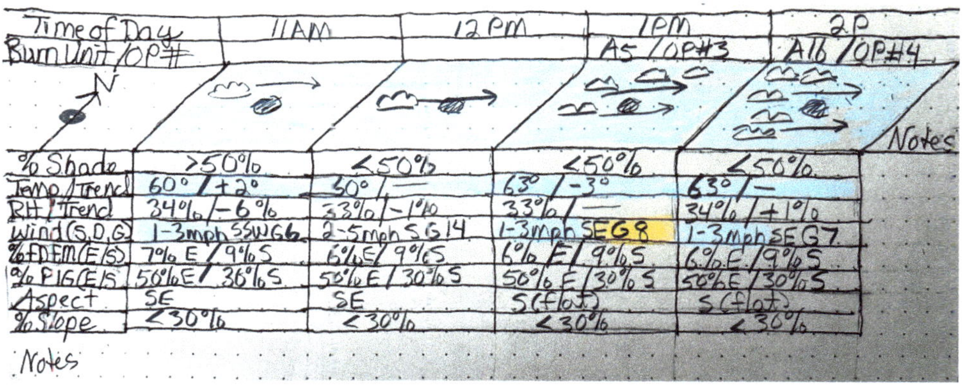

A few field tips for hourly weather observations:

- Take measurements just before the top of each hour so they align with operational weather updates.
- Record the observation location clearly (for example, Observation Point 3, northeast corner of Unit A14, or meadow near the access road).
- Sample conditions representative of the active fire or expected fire spread—not directly within flames or dense smoke.
- Note your elevational position relative to the fire (above, level with, or below the active burning area).
- Record canopy cover, cloud cover, slope, and aspect if they are not already documented for that Observation Point.
- Measure temperature and relative humidity in the shade, or with your back to the sun, using an appropriate weather instrument.
- Measure wind while facing into the wind.
- Add Fine Dead Fuel Moisture (FDFM) and Probability of Ignition (PIG) observations, discussed in the next chapter

Ecological Temperature & Humidity Indicators

Weather instruments tell us what the atmosphere is doing, while living organisms often reveal how those conditions are experienced within the landscape. As temperatures warm and sunlight reaches the vegetation, butterflies become more active, dragonflies begin patrolling open areas, bees increase their foraging, and many other insects respond to changing temperatures and humidity. Rather than treating insects as precise weather instruments, use them as ecological prompts for observation.

Fuel Moisture

The relationship between weather and vegetation is not merely a technical interaction; it is a continual exchange of energy that shapes the readiness of a landscape to burn. Morning temperatures and higher relative humidity slow drying and moderate fire potential. As the day progresses, solar heating, wind, and lower relative humidity gradually remove moisture from fine fuels, increasing their ability to ignite and carry fire. Fuel is not a static backdrop to a prescribed burn—it is an active participant in an ever-changing environment.

Observing these subtle changes transforms measurements into understanding. Through repeated journaling, weather, vegetation, and fuel moisture become connected rather than isolated observations. Over time, temperature, relative humidity, wind, fuel moisture, and your sensory observations begin telling the same story about how a landscape responds before, during, and after fire.

This chapter bridges the Vegetation and Weather chapters by focusing on the interaction between fuels and atmospheric conditions. It also prepares you for the Fire Ignitions, Behavior, and Smoke chapter by exploring how changing fuel moisture influences ignition potential, fire behavior, and the ability of a prescribed burn to meet its goals and objectives.

Begin by reviewing the weather forecast and prescribed burn parameters, then compare those expectations with what you observe in the field. As temperature, relative humidity, wind, canopy cover, and cloud cover change throughout the day, notice how those changes are reflected in the fuels around you. Fuel moisture measurements become most meaningful when interpreted within the context of the burn parameters. Likewise, sensory observations—how fuels feel, sound, smell, bend, or snap—become more valuable when they are repeatedly compared with measured weather conditions, Fine Dead Fuel Moisture (FDFM), and Probability of Ignition (PIG). Together, these comparisons help calibrate your observations and build a personal reference library that strengthens both the Study and Situational Awareness approaches.

Fine Dead Fuel Moisture

The primary focus of this chapter is **Fine Dead Fuel Moisture (FDFM)** because it responds rapidly to changing weather conditions and strongly influences ignition potential, fire behavior, and prescribed fire operations. Fine dead fuels correspond to the **1-hour fuels** introduced in the Vegetation chapter.

Estimating or measuring FDFM requires observations from both the Vegetation and Weather chapters, including:

- Air temperature
- Relative humidity
- Combined canopy and cloud cover (fuel shading)
- Slope
- Slope aspect
- Representative fuel model and Observation Point

FDFM Hack

Divide the relative humidity by 5 to estimate the FDFM.

These observations describe the conditions experienced by the fuels rather than simply the atmosphere above them.

FDFM is commonly used to determine when to begin ignition, adjust firing patterns, anticipate changes in fire behavior, and evaluate whether burn objectives can be achieved safely. For journaling purposes, interpret FDFM within the prescribed burn parameters whenever possible rather than relying on a single value alone. Comparing measured values with your sensory observations helps develop an intuitive understanding of changing fuel conditions.

General interpretation ranges include:

- **Less than 3–6%** — High fire intensity potential
- **7–15%** — Moderate fire intensity potential
- **Greater than 20%** — Low fire intensity potential

As discussed in the Weather chapter, FDFM can be added to your hourly weather tracker. When available, use NWCG field guides or electronic tools such as the Wildland Toolkit app to calculate FDFM. For journaling purposes, a simple estimate can often be made by dividing the relative humidity by five. Although less precise, this approximation provides another way to relate changing weather conditions to changing fuel moisture in the field.

Probability of Ignition

Fine Dead Fuel Moisture becomes even more meaningful when paired with **Probability of Ignition (PIG)**. PIG estimates the likelihood (0–100%) that a firebrand landing in receptive fine fuels will produce a successful ignition. Together, FDFM and PIG provide operational context for prescribed burning by linking weather conditions and fuel moisture to ignition potential and expected fire behavior.

PIG Hack

FDFM<5%=>90% PIG
FDFM 6-8%=60-90% PIG
FDFM 9-11%=30-60% PIG
FDFM 12-15%=10-30% PIG
FDFM >15%=<10%PIG

General interpretation ranges include:

- **80–100%** – Very high
- **50–80%** – Moderate to high
- **20–50%** – Moderate
- **Less than 20%** – Low

When journaling, record FDFM and PIG alongside your weather observations and compare them with the prescribed burn parameters. The reference table on the following page provides generalized FDFM and PIG ranges for common fuel model groups. These values synthesize information from NWCG training materials, USDA Forest Service fuel model references, the National Fire Danger Rating System, and prescribed burning guidance. They are intended to support observation, comparison, and journaling rather than replace local burn prescriptions.

Observing Fuel Moisture

Fuel moisture measurements provide valuable information, but they become far more meaningful when paired with direct observation. By touching, smelling, listening to, and gently testing fuels under different weather conditions, you begin connecting burn parameters, weather measurements, FDFM, and PIG with the way fuels actually behave in the field. These repeated comparisons help calibrate your senses and strengthen your ability to recognize changing fuel conditions during future prescribed burns.

Common FDFM & PIG Parameters by Fuel Model

Grass

Objective	Low	Moderate	High
FDFM	>15%	7-15%	<7%
PIG	<70%	70-85%	>85%

Grass-Shrub

Objective	Low	Moderate	High
FDFM	>15%	7-15%	<7%
PIG	<60%	60-75%	>75%

Objective	Low	Moderate	High
FDFM	>10%	6-10%	<8%
PIG	<35%	35-55%	>55%

Timber Understory (TU)

Objective	Low	Moderate	High
FDFM	>10%	6-10%	<8%
PIG	<50%	50-65%	>65%

Timber Litter (TL)

Objective	Low	Moderate	High
FDFM	>15%	7-15%	<7%
PIG	<40%	40-60%	>60%

Slash/Blowdown (SB)

Objective	Low	Moderate	High
FDFM	>15%	10-15%	<10%
PIG	<50%	50-70%%	>70%

As you move through the burn unit, notice how fuel moisture varies between exposed and shaded locations, across different fuel models, and throughout the day. Dry grasses may feel brittle and sharp, snapping cleanly between your fingers, while slightly moister fuels bend before breaking. Underfoot, dry litter often produces crisp cracks and hollow sounds, whereas higher fuel moisture softens your steps, creating a quieter, more cushioned feel. Continue building on the sound observations introduced in the fuel model dance and music exercise

from the Vegetation chapter. Notice how the rhythm, volume, and character of sounds change as fuels become drier or moister. Those same visual and musical symbols can become another way to represent changing fuel moisture.

Scent changes as well. Drier conditions often carry dusty, sun-warmed, or resinous aromas, while increasing moisture brings forward richer earthy or vegetal scents. These differences are often subtle, becoming easier to recognize through repeated observation and comparison with weather measurements and fuel moisture data.

Rather than relying on any single observation, combine what you measure with what you experience. Compare your sensory observations with the burn parameters, weather measurements, FDFM, and PIG throughout the burn day. As your observations accumulate, the relationship between weather, vegetation, and fuel moisture becomes more intuitive. Over time, your senses become calibrated to the burn parameters, allowing you to recognize when fuels are approaching, within, or moving beyond the desired operating range.

Snap Test

The snap test is a simple field exercise that connects fuel moisture measurements with direct sensory observation. Rather than replacing FDFM or PIG, it helps you recognize how changing fuel moisture feels, sounds, and responds as fuels move toward, within, or beyond the prescribed burn parameters.

Practice the snap test repeatedly under different weather conditions, fuel models, and burn windows. Compare your observations with measured weather conditions, FDFM, and PIG to gradually calibrate your senses and strengthen situational awareness.

What

Use naturally cured fine dead fuels such as pine needles, grasses, herbaceous stems, or fallen leaves.

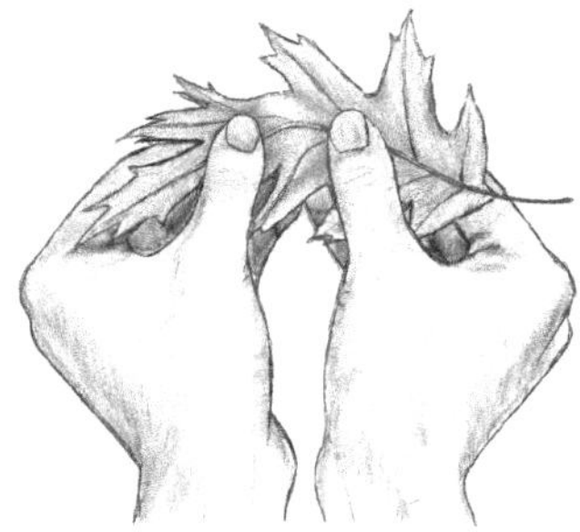

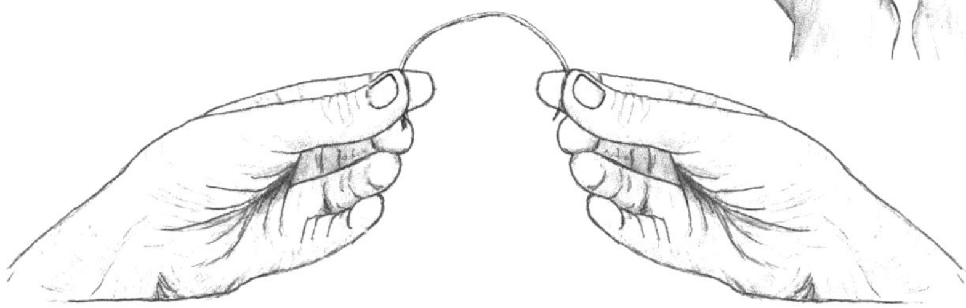

Where

Compare fuels from both exposed (sunlit) and shaded locations within representative Observation Points and fuel models.

When

Conduct the snap test during different burn parameter conditions throughout the day. Whenever possible, compare observations during cooler morning conditions, the preferred burn window, and the warmer afternoon period. Perform the test under lighter winds (approximately 3–8 mph) when practical, as stronger winds can create the sensation that fuels are drier than they actually are.

How

Hold the fuel at both ends and bend it slowly.

Notice:

- How far the fuel bends before breaking.
- Whether it breaks suddenly or gradually.
- The sound produced as it breaks.
- Whether leaf veins remain intact or fracture.
- Differences between exposed and shaded fuels.
- Differences among fuel models and plant materials.

Compare these observations with the current weather, burn parameters, FDFM, and PIG.

Snap Test Cues with FDFM & PIG Associations

•**Breaks instantly (crisp snap)** — FDFM <5% · PIG 80–100%. Common in grass fuel models.

•**Breaks on a wide bend** — FDFM 5–10% · PIG 50–80%. Common in grass, grass-shrub, and some slash/ blowdown fuel models.

•**Bends, then breaks on a narrow bend** — FDFM 10–15% · PIG 20–50%. Common in timber litter fuel models.

•**Bends fully without breaking** — FDFM >20% · PIG <20%. Often marginal for prescribed burning across many fuel models.

These relationships are general field observations rather than precise measurements. Fuel species, fuel bed structure, curing, recent weather, and microclimate all influence how fuels respond. Elevated or fluffy fuel beds, such as curled leaves or suspended grasses, often dry more quickly and may ignite more readily than flattened fuels experiencing the same weather conditions.

Use the symbolic bend and break sketches shown below to record your observations quickly in the field. Pair each symbol with a few descriptive words about the feel, sound, and fuel type, then add the corresponding weather measurements, burn parameters, FDFM, and PIG. The example journal page demonstrates one way to organize these observations and compare them throughout the burn day.

Over time, these repeated observations become another way to recognize changing fuel moisture before the measurements are even recorded. The goal is not to estimate exact moisture values from the snap test, but to strengthen your ability to relate sensory observations to burn parameters through repeated comparison and practice.

Burn Name Location Observer	**Activity & Observation Summary**	Date & Time Period of Activity

Description of one lesson learned to carry over to the next burn observation event/period.

Time	Major Events/Observations
10:45 A	**Snap Test**—in timber litter fuel model with pine needles in direct sun and shade. **Exposed fuels**—Bends, breaks on narrow loop Estimated **FDFM 10-15%, PIG 20-50%** **Shaded fuels**—Bends with no breaking Estimated **FDFM >20%, PIG <20%**

Live Fuel Moisture & Vapor Pressure Deficit

Unlike dead fuel moisture, live fuel moisture is traditionally measured by collecting vegetation samples, weighing them before and after oven drying, and calculating the amount of water contained within plant tissues. These methods are commonly used in fire planning and research but fall outside the scope of this journaling guide. Instead, this section focuses on observing living vegetation and relating those observations to Vapor Pressure Deficit (VPD).

VPD describes the atmosphere's drying demand—how strongly the air draws moisture from soils, living vegetation, and other surfaces. I like to think of VPD as a measure of how a living plant experiences changing weather conditions. Rather than describing the amount of water within the plant, VPD reflects the stress placed upon it by the atmosphere. As VPD increases, plants lose moisture more rapidly through transpiration and begin conserving water by partially closing their stomata. Although this reduces water loss, it also slows photosynthesis, growth, and other physiological processes.

VPD is determined by both temperature and relative humidity, but temperature generally has the stronger influence because warm air can hold much more water vapor than cool air. As temperatures rise without a corresponding increase in atmospheric moisture, VPD increases and plants experience greater moisture stress. For this reason, VPD often provides a better indication of plant stress than temperature or relative humidity considered independently.

The atmosphere's drying demand can also be experienced personally. During periods of high VPD, many people notice dry skin, chapped lips, dry eyes, or increased thirst. Although plants respond differently than people, these shared experiences provide another way to relate to changing atmospheric conditions.

The visual chart that follows relates general VPD ranges to subtle observations of trees, shrubs, and herbaceous vegetation. Look for gradual changes in leaf or needle color, texture, flexibility, posture, angle, curling, and overall appearance. Compare several characteristics together rather than relying on a single observation.

Like the bird, butterfly, and other ecological indicators introduced throughout this book, these vegetation observations are intended to strengthen relationships rather than replace measurements. Compare them with weather, VPD, season, recent precipitation, canopy cover, slope, aspect, and other microclimate conditions. Revisiting the same plants throughout the burn season builds a personal observation library that helps distinguish normal seasonal development from increasing moisture stress.

Leaf/Needle Indicators of Vapor Pressure Deficit

	Low VPD 0-0.8 kPa	**Moderate VPD** 0.8-1.6 kPa	**High VPD** >1.6 kPa
Conifer Needles	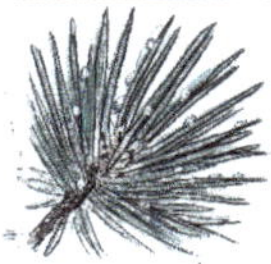 • Needles dark green, glossy, and plump appearance • Cool and pliable to touch • Branch tips upright	• Needles healthy green, normal spacing. Flexible, with slight resin tack to touch • Branch tips slightly upward or outward and Canopy appears full 	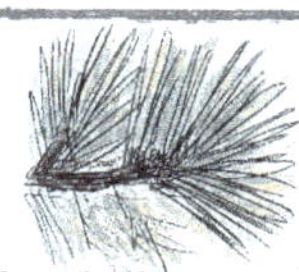 • Needles dull green, bluish, or grey-green, with slight curl or twist • Branch tips droop • Needles warm, dry with sticky or dry tack to touch
Broad Leaves	 • Leaves open and flat; dark saturated green • Leaves cool to touch, soft and supple to touch • Canopy full and lush	• Leaves open and tops mostly oriented to sun, moderate green • Leaves slightly cool, firm, flexible • Canopy appears moderate density 	 • Leaves lighter in color, cupping/ curved or limp. Sometimes oriented sideways • Leaves warm, dry to touch • Extreme cases leaves dry/ scorched from heat
Small Leaves	 • Leaves full, green, shiny and turgid • Leaves cool, soft and supple to touch • Stems somewhat flexible	• Leaves moderate green and mostly oriented to the sun • Leaves feel slightly cool, firm, and slightly pliable 	 • Leaves dull green, edges curled, and/or angled sideways • Leaves feel warm, leathery, less flexible • Twigs stiff
Grass Blades	 • Blades stand tall, wide, open and flat, lush green • Dew common • Blades cool, soft, and flexible to touch	• Blades normal green and standing upright • Blades feel slightly cool, firm and springy 	 • Blades dull green, with brown tips curled/rolled inward • Blades warm, dry, and less flexible

Building an Observation Reference Library

The observations in this chapter are intended to be repeated, compared, and refined over time. Consistent journal formats make it easier to recognize patterns across burn periods, seasons, fuel models, and prescribed burns. Rather than collecting isolated notes, you begin building a personal reference library that strengthens both the Study and Situational Awareness approaches.

This reference library also brings together the observations from the previous chapters. Landforms shape terrain and microclimates. Vegetation influences fuel structure and continuity. Weather changes throughout the day. Fuel moisture reflects the ongoing interaction between weather and vegetation. No single measurement or observation tells the whole story.

Fuel Model Reference Library Exercise

Select a single fuel model and revisit it throughout the burn season whenever possible. Rather than trying to document everything during one visit, allow your observations to grow over time. Each visit adds another layer of understanding.

Step 1. Record the Setting

Begin by recording the date, time, season, burn unit, Observation Point, and fuel model. If helpful, include a simple sketch or reminder of the vegetation structure and surrounding landscape.

Step 2. Record Burn Parameters

Add the current burn parameters whenever they are available, including temperature, relative humidity, wind speed and direction, FDFM, and PIG.

Step 3. Record Fuel Moisture Observations

Add observations related to both dead and live fuels. Include snap-test observations, VPD, changes in living vegetation, canopy and cloud cover, and any weather measurements that help explain changing fuel conditions.

Step 4. Add Sensory Observations

Record how the landscape feels, sounds, smells, and responds as conditions change. You may also include ecological observations that deepen your understanding of place.

Step 5. Revisit the Same Location

Return to the same Observation Point before, during, and after prescribed burns whenever possible. Compare observations across different weather conditions, burn windows, and seasons. Resist the temptation to begin a new page each time. Instead, allow the page to evolve as your understanding of that fuel model grows.

Montane Timber Litter & Adjacent Timber Understory Optimal Burn Parameter Sensory Observations

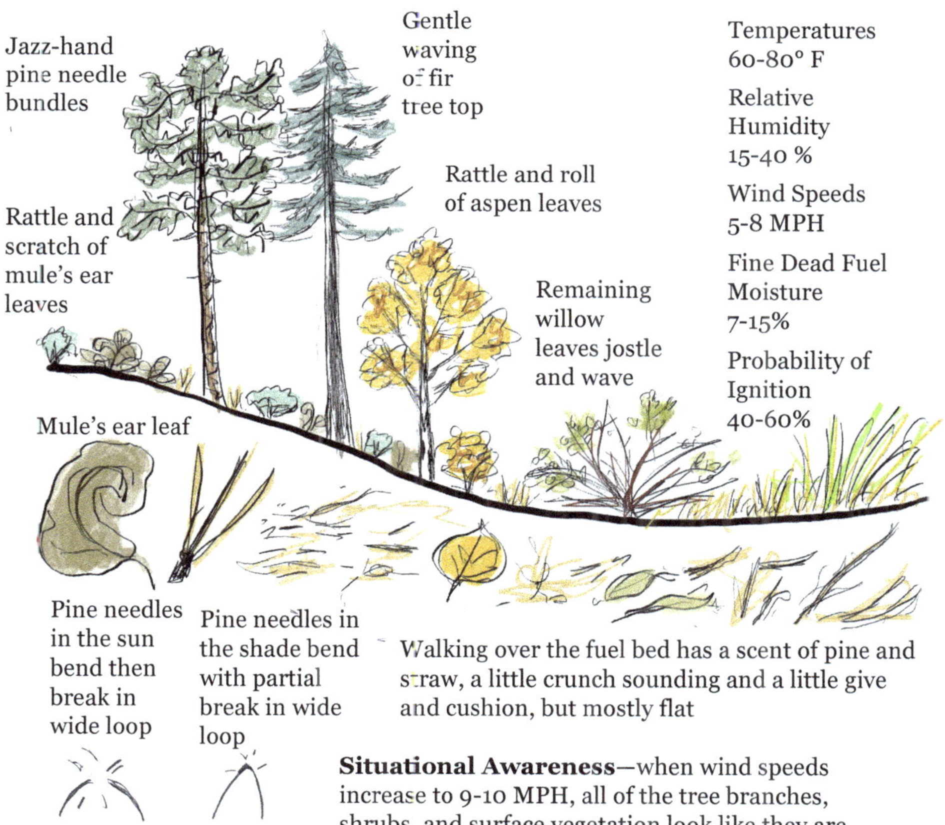

Situational Awareness—when wind speeds increase to 9-10 MPH, all of the tree branches, shrubs, and surface vegetation look like they are being shoved rather than waving and pulsing. The sounds mix and become hard to differentiate.

Observing Relationships

Fire is often described through its individual parts—weather, vegetation, terrain, moisture, smoke, flame, and the people who work with it. Yet none of these exists in isolation. Fire is an environmental process, continually shaped by relationships among people, place, weather, vegetation, and time. As you have worked through the previous chapters, you have already begun observing these individual elements. Now it is time to begin observing the relationships that connect them.

Observation is more than noticing what surrounds us. It is also noticing how our understanding changes, what captures our curiosity, where questions emerge, and how our own relationship with the landscape continues to deepen. Journaling creates the space to slow down, reflect, and organize those observations into something more enduring than a moment in time.

Thinking Through Relationships

Traditional nature journaling often encourages the use of three familiar thinking prompts: **I Notice... I Wonder... It Reminds Me Of...** These prompts cultivate curiosity, encourage careful observation, and invite reflection without requiring specialized knowledge. They remain an important part of my own journaling practice and continue to shape how I approach every landscape.

Pyrosketchology builds upon these familiar prompts by expanding their use to observe and reflect upon the relationships that shape the fire environment. Rather than observing individual features in isolation, we begin exploring how people, place, weather, vegetation, fire, and stewardship influence one another. The prompts become a way to organize our thinking as much as our observations, helping us connect measurements with sensory observations, individual observations with larger environmental processes, and ourselves with the landscapes we seek to understand and care for.

As you work through the remainder of this book, consider these prompts as a repeating cycle rather than a checklist. You may move back and forth among them many times while observing a single burn or revisiting the same place throughout the year.

I Notice...

Purposefully observe the influential elements and relationships within the fire environment using words, numbers, and pictures.

- *I notice* the relationship between the ignition tools being used and the backgrounds, experience, and stewardship traditions of the burn partners.
- *I notice* where operational decisions or ignition techniques change as conditions change.
- *I notice* details that surprise me, repeat themselves, or invite a closer look.

I Wonder...

Transform observations into questions that deepen understanding.

- *I wonder why* the burn team chose this ignition pattern instead of another.
- *I wonder how* today's weather and fuel conditions are helping achieve the burn objectives.
- *I wonder what* additional observations or measurements would help me better understand these relationships.

It Reminds Me Of...

Use analogy, comparison, and prior experience to understand larger relationships.

- *It reminds me of* a dance, where each practitioner adjusts their movements in response to changing weather, fuels, and one another.
- *It reminds me of* an orchestra, where different fuels respond in their own way while together creating the larger composition of fire behavior.
- *It reminds me of* another burn, season, or landscape where similar conditions produced different outcomes.
- *It reminds me that* my journal is not simply recording observations—it is helping me discover relationships between people, place, and fire that I might otherwise overlook.

Ignitions & Holding

Everything you've observed in the previous chapters now begins to converge. Landforms shape the terrain and its microclimates. Vegetation influences how fire can move through the landscape. Weather changes throughout the day, and fuel moisture reflects the ongoing dialogue between the atmosphere and living and dead vegetation. Together, these observations create the conditions that determine how, when, and where fire can be introduced.

This chapter is where those relationships become actions. Rather than simply observing the fire environment, we begin observing how practitioners intentionally work within those relationships. Through the thoughtful selection of tools, ignition techniques, patterns, timing, and holding operations, they guide fire toward desired ecological and cultural outcomes.

Unlike wildfire, prescribed and cultural burning begin with intention. Fire does not simply appear across the landscape—it is thoughtfully introduced through a series of intentional decisions. Burn goals establish the purpose. Burn objectives define the desired outcomes. Burn parameters define the environmental conditions within which those objectives can be safely and effectively pursued. Within that framework, the burn team continually adapts ignition strategies as conditions change throughout the day.

Forecasts help shape the ignition plan before the burn begins, while observed weather, fuels, and burn parameters continually inform decisions throughout the burn. Burns may begin later than planned because fuels remain too moist, ignition spacing may be adjusted as weather evolves, or operations may pause until conditions once again meet the desired parameters. Sometimes all objectives are achieved. At other times, only some objectives are accomplished because changing conditions require practitioners to adapt to the opportunities the landscape provides. These adjustments are not failures of the burn plan—they are examples of experienced practitioners thoughtfully responding to changing conditions.

As you observe the burn, notice which tools are being used, how fire is introduced to the ground, where ignition patterns change across the landscape, and how those changes relate to terrain, vegetation, weather, burn objectives, and burn parameters. These observations help reveal the relationships between practitioner decisions and the fire's response, providing the context for understanding the fire behavior explored in the following chapter.

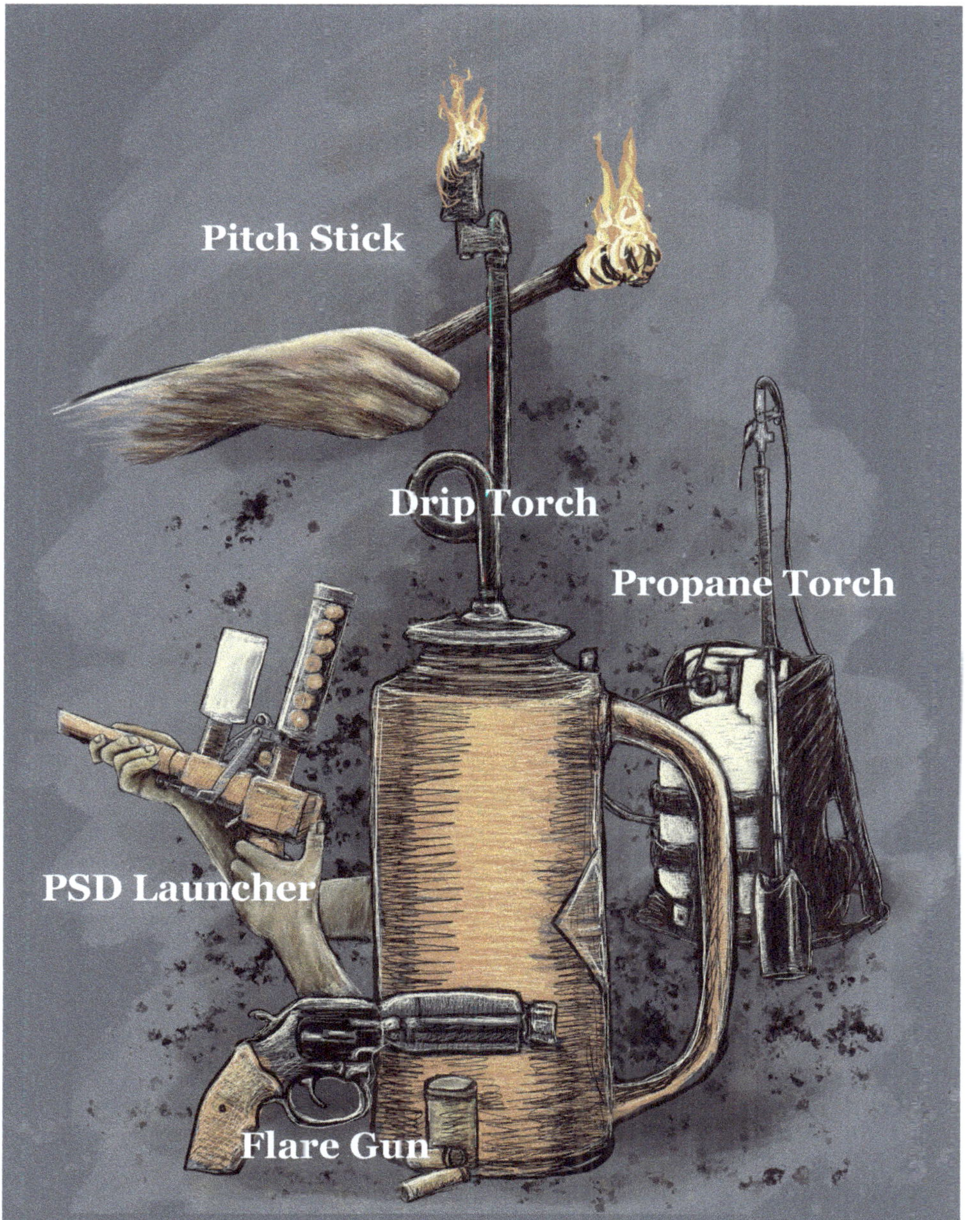

It Reminds Me Of...

As I watch ignitions unfold, I'm often reminded of a dance. The landform becomes the stage. Weather becomes the music. Ignition techniques and holding operations become the choreography. Fire behavior becomes the dancer, responding to every change in rhythm, movement, and place. Like all analogies, this one isn't meant to explain every part of fire. Instead, it reminds me that good fire emerges from the ongoing relationship among people, place, and fire.

Preparing to Journal the Burn

No two prescribed or cultural burns unfold exactly the same way, and neither should your journal. Before a burn begins, I prepare journal pages that give me options rather than obligations. I almost always include an hourly weather tracker, even if I suspect I'll only complete a few observations. I may also prepare a burn unit map, pages for mapping ignition patterns, quick sketch pages, or layouts for Story, Study, and Situational Awareness observations. Some pages may remain nearly blank, while others become filled with notes, sketches, questions, and reflections.

I let the burn determine which observations become most important.

The journal pages related to those on the next page, included a weather forecast table using the burn parameter color codes described earlier in the book and included an hourly weather tracker so I would have the option of using a Study journaling approach if the situation warranted. Once I arrived on site, however, I realized the burn unit was relatively small and only a single ignition pattern was being used. Rather than creating a burn unit map, I decided a larger sketch showing the oak woodland, the igniters slowly pulling the ignition line downslope, and the overall setting would better capture the burn.

Because this was a fully traditional cultural burn, I naturally shifted toward a Story approach. I wanted to preserve some of the memorable moments and cultural context, so I included a mix of related sketches and small photographs. One of the burn objectives was to maintain an acorn gathering area, so I also added a sprouting acorn and an oak leaf as reminders of the relationship between fire, oak stewardship, and future harvests.

This example illustrates an important aspect of good fire journaling. Preparing several journal formats before the burn gives you options, but the burn itself determines which observations become most meaningful. The templates continue to organize measurements, observations, and comparisons throughout the burn, but the emphasis may shift. Some burns reveal strong patterns through a Study approach, while others naturally lead toward Story or Situational Awareness. The prepared pages continue to organize your observations while giving you the flexibility to emphasize the relationships that become most important.

Some burns invite careful study of ignition techniques and fire patterns. Others become stories about stewardship, conversations, and the people working together on the landscape. On more dynamic burn days, Situational Awareness observations and hourly weather changes naturally become the priority. You may also decide before the burn to focus on a single journaling approach based on your interests, available time, or experience. Preparing additional journal formats simply gives you the flexibility to shift your emphasis if the burn unfolds differently than expected. Even when a Study or Situational Awareness template is not fully developed, it continues to serve as an observation guide, reminding you which measurements and relationships are most relevant to understanding the fire environment.

This approach mirrors prescribed fire itself. Burn plans, forecasts, and burn parameters provide structure, while experienced practitioners continually adapt their ignition strategies as weather, fuels, and fire behavior change. Journaling follows a similar process. Before the burn, set your intentions if you can. Prepare the observation tools, supplies, and journal formats that best support those intentions, if time allows. During the burn, let your observations guide how you use those prepared pages. Some templates may remain only lightly completed, while others become filled with notes, measurements, sketches, and reflections. Preparation adds flexibility, allowing your journal to adapt to the burn while maintaining the organization needed to recognize patterns, relationships, and memorable experiences.

As the burn begins to unfold, notice not only which tools are being used, but also how, when, and where fire is introduced to the landscape. Observe how ignition patterns change across different fuels and terrain, how holding operations respond, and how practitioner decisions adapt to changing weather and burn parameters. These observations provide the context for understanding the fire behavior explored in the next chapter.

Observing Ignition Tools

Ignition tools vary widely among prescribed and cultural burns, reflecting differences in burn objectives, traditions, fuels, and the landscapes in which they are used. Some burns rely primarily on drip torches or propane torches, while others may incorporate traditional methods such as friction fire, ember nests, pitch sticks, or other culturally specific practices. Regardless of the tools themselves, the more meaningful observations often come from how and why they are used.

When journaling, look beyond simply identifying the equipment. Notice who is using each tool, how it is prepared, when it is used, and what it reveals about the relationship between people, place, and fire. The tools themselves often tell a story about stewardship, traditions, burn objectives, available resources, and the landscape being cared for.

My following journal page highlights observations from the same cultural burn introduced on the previous page. Rather than simply recording the ignition tools, I combined photographs, sketches, and notes to document how the tools were prepared and used. Most of the ignition relied on traditional tribal methods, although a propane torch was used in a few locations where it supported the burn objectives while minimizing petroleum residue that could affect the value and use of harvested acorns. The tools and the ways they were used carried much deeper meaning than I typically experience during prescribed burns and were well worth taking the time to journal.

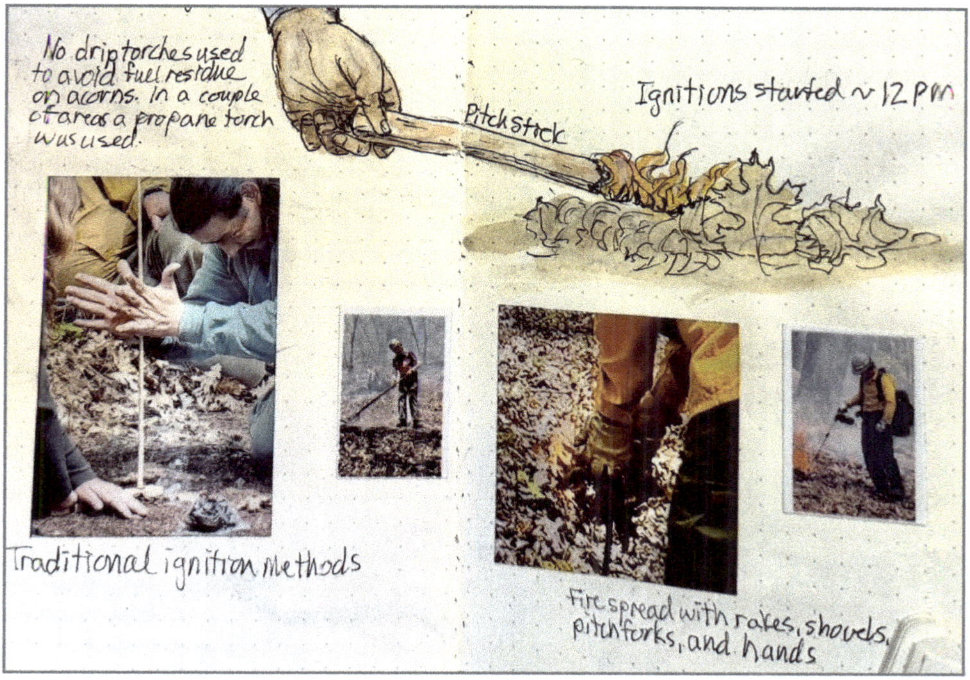

Observing the Test Fire

The operational test fire marks the transition from observing environmental conditions to observing the relationship between those conditions and the fire's response. Until this point, your observations have focused on understanding the landscape and determining whether conditions are likely to support the desired fire behavior. The test fire is the first opportunity to compare those observations with the fire's actual response. It provides a baseline for comparison throughout the remainder of the burn, helping practitioners evaluate whether fuels are receptive, weather conditions remain within the desired burn parameters, and initial fire behavior aligns with burn goals and objectives.

Before a prescribed burn begins, the ignitions team conducts a test fire to determine whether fire will carry through the fuels as expected and whether fire behavior is appropriate for the planned operations. If conditions are not within the desired burn parameters, ignitions may be delayed, modified, or postponed altogether. The test fire is one of the first opportunities to compare forecasts, burn parameters, and field observations with the fire's actual response, helping guide decisions throughout the burn.

Use the following to help frame your observations:

- Record the **time, burn unit, Observation Point (OP),** and current weather conditions.
- Note where the test fire is placed and why that location may have been selected.
- Observe how readily the fuels ignite, whether the fire carries through the fuel bed, the amount of heat produced, initial flame characteristics, and early smoke development.
- Compare these observations with the fuel moisture, probability of ignition, weather conditions, and other observations you recorded before the burn.
- Use the test fire as your baseline. As the burn progresses, return to these initial observations to compare how changing weather, fuels, and ignition patterns influence fire behavior throughout the day.

Leaf Ignition Exercise

This exercise is not intended to replace the operational test fire or estimate fuel moisture. Instead, it helps strengthen the relationship between your sensory observations, burn parameters, fuel moisture, and probability of ignition. This type of test fire may be used but is not a standard operational practice during most prescribed burns. To do this exercise, get permission from the burn team and determine where , when and how it can be safely conducted.

The illustrations and exercise following are adapted from guidance shared by Dr. Don Hankins (California State University, Chico) and the Butte County Fire Safe Council. Similar approaches may be used during cultural burns or, with approval from the burn team, as a simple learning exercise to better understand how fine fuels respond to ignition.

Select a single cured leaf from the ground in an unshaded area for each leaf ignition position.

Follow the instructions and note observations in your journal.

Do not let the leaf burn completely and reach your fingers.

Too Moist for Ignition

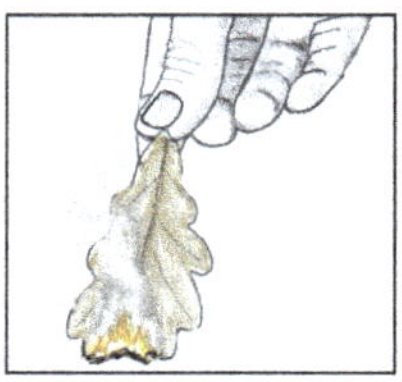

Hold a leaf upside down from the stem and ignite at the tip. If the leaf does not easily burn from the tip to the base, the fuels are too moist to start the burn.

Just Right for Some Burns

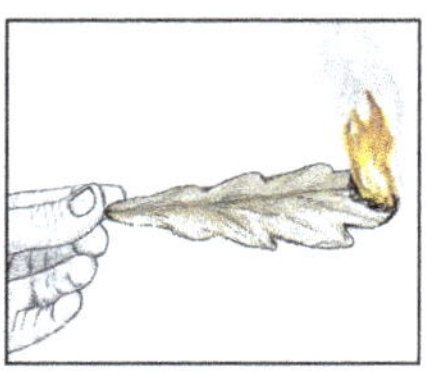

Hold a leaf horizontally from the stem and ignite at the tip. If the leaf easily burns from the tip to the base, the fuels are good for some burn parameters and not dry enough for some others.

Too Dry for Some Burns & Just Right for Others

Hold the upright from the stem and ignite it at the tip. If the leaf easy burns from the tip down, the fuels may be too dry for some prescriptions or ideal for others.

Fire Ignitions

Everything you've observed to this point now begins taking shape on the landscape. Ignition techniques and patterns are the first visible expressions of how practitioners intentionally work within the relationships you've been observing throughout this book. Rather than simply introducing fire, practitioners apply it in ways that reflect burn goals and objectives, burn parameters, terrain, fuels, weather, and stewardship decisions. Wind direction and slope position are especially important because they determine where, when, and how different ignition patterns are applied. As wind direction changes or ignitions move to different positions on the slope, practitioners often adjust ignition patterns to maintain the desired fire behavior and burn conditions. These observations provide the context for understanding the fire behavior discussed in the next chapter.

> ***I Notice...*** where and when different ignition techniques and patterns are used across the landscape.
>
> ***I Wonder...*** which terrain, fuels, weather, burn parameters, and burn objectives may have influenced those decisions.
>
> ***It Reminds Me Of...*** how practitioners work with the landscape rather than against it, adapting fire to the place instead of forcing the place to fit the fire.

It may seem counterintuitive, but techniques that place more fire on the ground often result in lower overall fire intensity. I describe ignition techniques and patterns as **"leashed" fire**. Once fire spreads beyond its initial ignition placement, it becomes **"loosed" fire**, which I use for framing fire behavior observations. In general, more controlled ignition placement leads to more moderated fire behavior. The illustrations on the following page provide a visual comparison of the fire ignition techniques.

Fire Ignition Techniques

The first step is learning to recognize the different ways fire is intentionally applied to the landscape.

Ignition techniques are often discussed together with ignition patterns, but separating them helps clarify observation. In this text, ignition techniques refer specifically to **how** fire is applied to the ground, while ignition patterns describe **where** and in what sequence those techniques are applied across the burn unit. These approaches are frequently combined and adjusted to meet burn goals.

Ignition Lines

Typically created with a drip torch, ignition lines can vary in intensity based on spacing. Wider spacing generally produces higher fire intensity. This technique is commonly used for fuel reduction and invasive species management.

Ignition Dots

Dots are placed in organized string patterns, allowing individual dots to merge gradually. Spacing influences intensity and flame variability. This approach often produces diverse ecological effects and is commonly used in habitat enhancement burning.

Ignition Points

Points are placed more irregularly and may be applied using a wider range of tools, including flare guns or PSD systems. Cultural burning may involve igniting specific species or age classes. Fire intensity varies depending on fuels and weather.

Ignition Painting / Fill

This technique involves splashed and widely distributed fire across an area, often resulting in lower fire intensity. It is commonly used to reduce fuels around sensitive features such as snags or legacy trees.

Terminology, Description & Observation

As ignition patterns become more varied, so do the words used to describe them. Learning the terminology used by the burn team is helpful because it provides a common language for communication, operational briefings, and documentation. These names help you recognize and discuss the ignition patterns being used during the burn.

At the same time, the terminology is only one way of understanding what you observe. Operational names describe how ignitions are organized across the landscape, while descriptive language often reflects the relationships among people, place, and fire. During a cultural burn, for example, you may hear fire described as being *introduced more gently*, *allowed to find its own path*, or *introduced in conversation with the wind*. Holding operations may be described as *protecting sensitive areas* rather than emphasizing the specific tactics being used.

As you journal, begin by learning and using the terminology defined by the burn team, especially when your journal may support operational reports or future documentation. Then consider adding your own descriptive observations that reflect what you notice about the relationships between the tools, techniques, terrain, weather, vegetation, and the people applying the fire. Both approaches strengthen your understanding. One provides a shared operational language; the other helps build deeper observation, connection, and reflection.

Perimeter Ignition Patterns

Ignition patterns describe the spatial placement of ignition techniques throughout the burn unit. Perimeter ignition patterns are those placed along various positions of the burn unit boundary, which serves as the operational control line during the burn. These are often used as the baseline for safe and effective ignitions on the interior. Ignition patterns influence fire intensity and behavior across different fuel models, slopes, and weather conditions (Martin & Hamman, 2016).

Wind direction and slope determine the positioning of head, flank, and back fire ignitions. These ignition line patterns may shift if wind direction changes during the burn. Maintaining a weather forecast and hourly tracker guides and informs ignition patterns.

Wind Aligned Perimeter Ignition Patterns

Note—the gray frame of the illustration represents the burn unit control lines and boundary.

Slope Aligned Perimeter Ignition Patterns

Back Fire Pattern (Baseline)

Initiated on the downwind/leeside or upslope perimeter moving against wind or downslope. Produces lower intensity fire with lower flame lengths and slower rates of spread but can increase fuel consumption and plant root impacts.

Flank Fire Pattern

Ignitions start from one or both sides of the control line perpendicular to the direction of wind or slope. Typically produces mixed fire behavior and often incorporates interior ignition strategies.

Head Fire Pattern

Initiated on the upwind/windward or downslope control line producing faster spread and longer flame lengths. Usually implemented after establishing a secure back fire ignition line.

Perimeter Wrapping/ Ring Fire Pattern

Combines back, flank, and head fire ignitions to move fire from the perimeter inward. Commonly used in grassland burns and often results in increasing intensity toward the final stages.

> **Beyond the Operational Name**
>
> Operational terminology provides a common language for burn operations. Your journal can also include descriptive observations that reflect the relationships you notice or hear others describe.
>
> **Back Fire (Baseline)**
>
> - Fire introduced gently against the wind.
> - Fire walking slowly into the landscape.
>
> **Flank Fire**
>
> - Fire moving alongside the wind.
> - Fire introduced in conversation with the wind.
>
> **Head Fire**
>
> - Fire moving with the wind.
> - Fire released where the landscape is ready to carry it.

Interior Ignition Patterns

Once a secure perimeter ignition has established the operational control line, interior ignition patterns are used to continue introducing fire across the burn unit while moderating fire behavior and meeting burn objectives.

Interior patterns build off and occur within the boundary ignition area and are defined by igniter spacing, sequencing, and timing. The examples I have included are common, but there are other fire ignition patterns that I have not included. Some of the examples I have may be called by a different name, so always ask the burn team for a description of the patterns being used if you are unfamiliar with that name. The terminology used on your burn may differ by agency, organization, or burn team.

Strip Distance & Stagger

Ignition lines within the interior of the burn unit build off the perimeter ignition lines and are used to moderate flame lengths, rate of spread, and fire intensity. These lines are laid out in strips with a certain distance between igniters, which frames the fire behavior within narrow strips of vegetation instead of the entire unit area. Strips are often parallel to each other but may be staggered to reduce risks to the igniters on the upwind and upslope side.

Typical strip distances between igniters (Andrews et al., 2011):

- Short grass: 20–60 ft
- Grass–shrub: 15–20 ft
- Shrub: 10–30 ft
- Timber litter: 10–25 ft
- Heavy slash: 5–15 ft

No Stagger

Igniters move simultaneously at similar spacing between burn strips.

Strip Stagger (1-2-3 and 3-2-1)

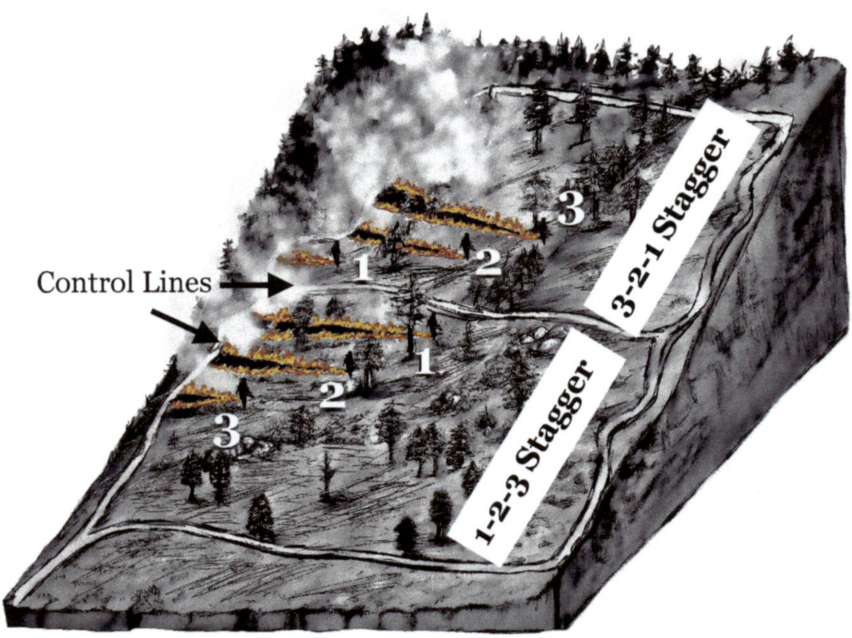

In a strip stagger, the person labeled as number one is always closest to the parallel control line to the direction of strip ignitions.

- In the **1-2-3 stagger,** the first igniter starts a few minutes before the next person and so on for the stagger between people. There can be more than three igniters and the sideways spacing between ignition lines can vary.
- In the **3-2-1 stagger**, the first person to start is the person labeled as number 3, then others follow in the timing stagger.

Strip-Head Fire Pattern

Ignition lines are at a right angle to the wind direction and slope, typically starting from the back fire area. This approach increases control by reducing the distance of head fire movement, burning only small strips of vegetation.

Strip-Flank Fire Pattern

Ignitions progress from the back fire burn line in strips parallel to the wind or slope. Strip width varies with fuel model and objectives. The image below is from my journal and shows strip flank ignition lines coming downslope on the burn unit map. No visual provided.

Zig-Zag Ignition Pattern

An igniter moves diagonally across the wind or down a slope, laying dots or short strips of fire at regular intervals while weaving back and forth across the unit. This pattern allows for quick coverage across large areas while ensuring the dots merge, creating a "junction zone." This pattern can modify fire behavior by changing the pace and distance between drops and the spacing between ignition lines.

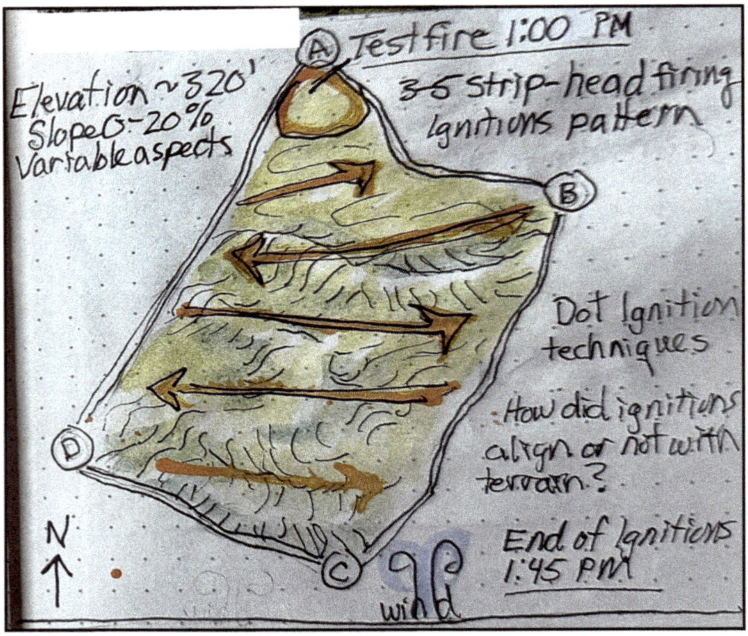

Wagon Wheel/Pivot Fire Pattern

Igniters use a 1-2-3 stagger along the top of the slope and control line for application of lines, then pivot and return to the starting direction along the black lines created in the preceding strip ignitions in another 1-2-3 stagger. The backline becomes the extended control line.

See the illustration on the top of the next page.

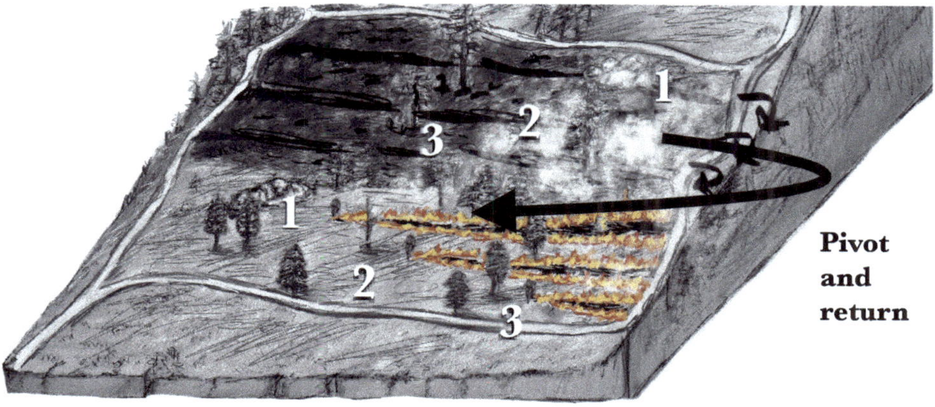

Terrain Change/Drainage Ignition Pattern

Ignition strips stop well before descending downslope into a drainage until another group of igniters can begin ignitions from the top of the opposite slope. Both ignition teams can then bring fire downslope toward the drainage.

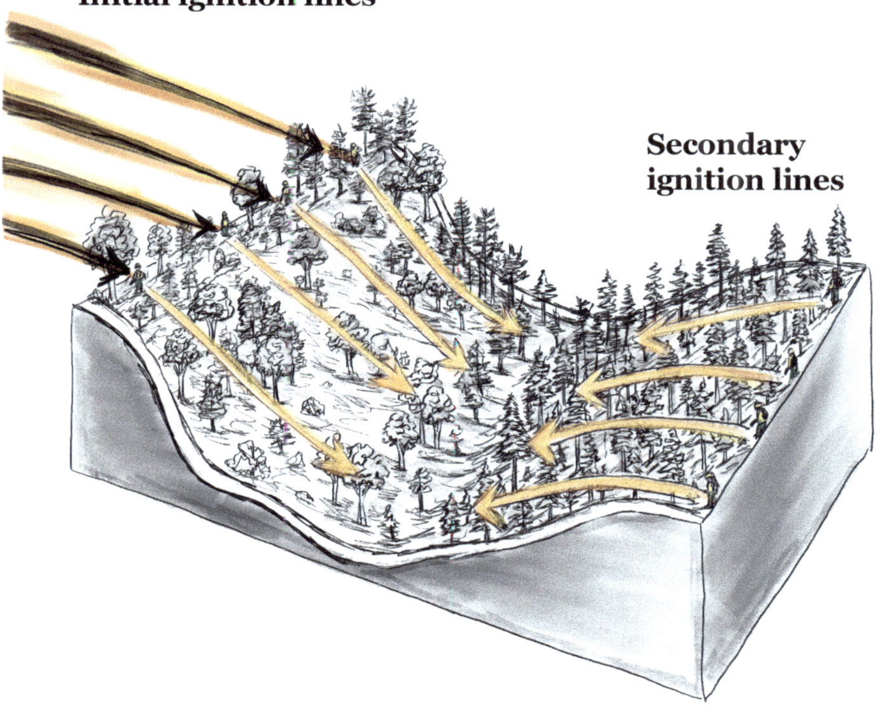

Interior Ring & Box Ignitions

Interior ring or box ignitions are used to protect sensitive features or manage fuels prior to broader ignition patterns. They may be placed around individual trees, stands of snags, concentrations of shrubs, piles of logs, buildings, cultural resources, or other sensitive features.

Unlike the larger perimeter ring fire pattern, which often allows fire to build intensity as it moves inward across the burn unit, interior ring and box ignitions limit the distance available for fire to spread, helping moderate fire behavior around the protected feature. Depending on burn objectives and conditions, the ring or box may not be completely closed, leaving the windward or downslope side open so the fire can slowly fill in from the edges rather than surrounding the feature with a continuous line of fire.

Ring ignitions are one approach used around homes and infrastructure prior to prescribed or wildfire flames can come from the outlying area. Interior ring and box ignitions are also used around things like a stand of dead trees, pile of logs, or contracted area of brush.

Tree Ring Ignition Pattern

Tree ring ignitions are a common type of interior ring ignition used to reduce surface fuels around individual trees. Ignition rings are typically placed **one to three feet from the base of the tree**, depending on tree size, bark characteristics, fuels, and burn objectives. This short ignition distance limits the distance available for fire to build flame length and vertical fire movement into the tree crown while still allowing fuels around the tree to be consumed.

Adaptive Ignition Decisions

The ignition techniques and patterns described in this chapter provide a common framework for introducing fire across a burn unit. Experienced practitioners continually adapt both the techniques they use and the way ignition patterns are applied in response to changing fuels, weather, terrain, season, burn objectives, and stewardship considerations. These adaptive decisions may not have formal names, but they often become some of the most interesting observations to capture in your journal.

For example, practitioners may intentionally introduce fire into either shaded or unshaded areas depending on burn objectives and conditions. Notice where these decisions occur across the landscape and consider how shading, fuels, weather, and desired fire behavior may have influenced those choices.

In grasslands, ignition decisions may also be adapted to seasonal curing. An initial pass may ignite only the driest patches of cured grasses, leaving greener vegetation unburned. As curing progresses or conditions change, a second pass can introduce fire into the remaining grasses. Previously burned patches help moderate fire spread during the second pass, allowing practitioners to work with changing fuel conditions rather than treating the grassland as a uniform fuel bed.

Ignition point techniques also provide opportunities for fine-scale adaptation. During some cultural burns, ignition points may be placed only on particular plant species or age classes to meet stewardship objectives. In other situations, dense vegetation or limited access may make it difficult or unsafe to carry ignition lines through the burn unit. Ignition points placed with a flare gun or plastic sphere dispenser (PSD) can establish small areas of fire that spread naturally toward surrounding ignition patterns or control lines.

The names of ignition techniques and patterns remain the same, but their application continually adapts to the relationships among fuels, weather, terrain, season, burn objectives, and stewardship considerations. These are often the observations that transform a journal from a record of what happened into an understanding of why practitioners chose to introduce fire in different ways across the landscape.

Journaling Ignition & Holding

The previous sections introduced the terminology, techniques, and ignition patterns commonly used during prescribed and cultural burns. The next few journal examples illustrate one way these observations can be organized in the field. Rather than focusing on a single journaling approach, they show how maps, hourly trackers, sketches, photographs, notes, and data visualizations can work together to document and study the relationships of ignition and holding operations with terrain, fuels, weather, and the resulting fire behavior.

A single journal spread rarely tells the complete story of a burn. Instead, each page captures one part of a larger collection of observations. As additional pages are added throughout the day, relationships begin to emerge between ignition decisions, weather, fuels, terrain, fire behavior, holding operations, and fire effects. Looking back across several journal pages often reveals patterns and insights that would be difficult to recognize from any one page alone.

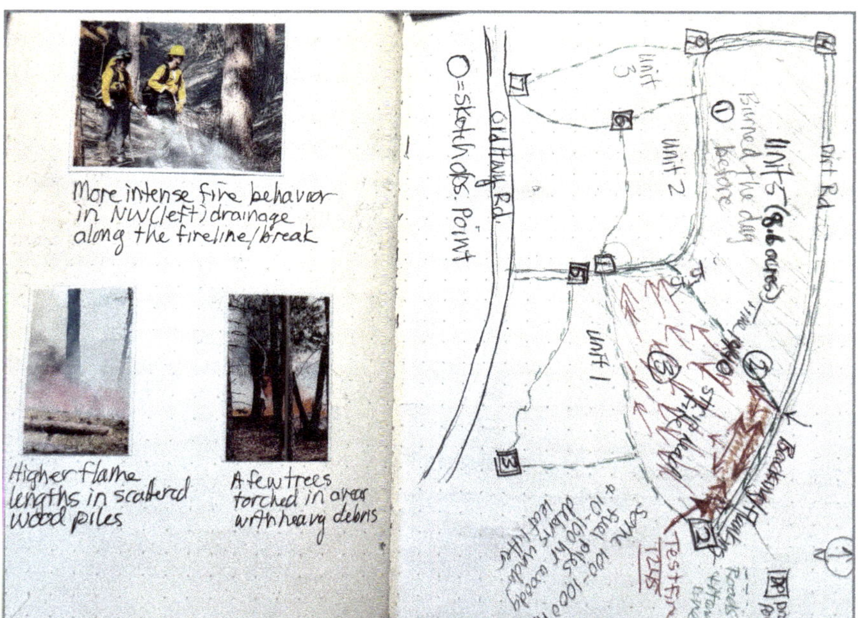

The above is a two-page spread combining a burn unit map with photographs and notes about ignition and holding activities. Although prepared as a Study journal, the combination of mapped ignitions, operational observations, and memorable moments also tells the story of the burn.

As you explore these examples, notice not only what is recorded, but also how the information is organized. Different layouts encourage different kinds of observations and reflections. Some pages naturally support storytelling, while others emphasize comparison, analysis, or situational awareness. Together, they illustrate how a collection of well-organized observations can become much more valuable than any single journal page by itself.

Small sketches integrated into an hourly observation tracker preserve ignition tools, techniques, and practitioner activities while maintaining a structured record of weather, location, and operational observations. Photographs taken in the field can be translated into sketches later, allowing additional details and reflections to be added after the burn.

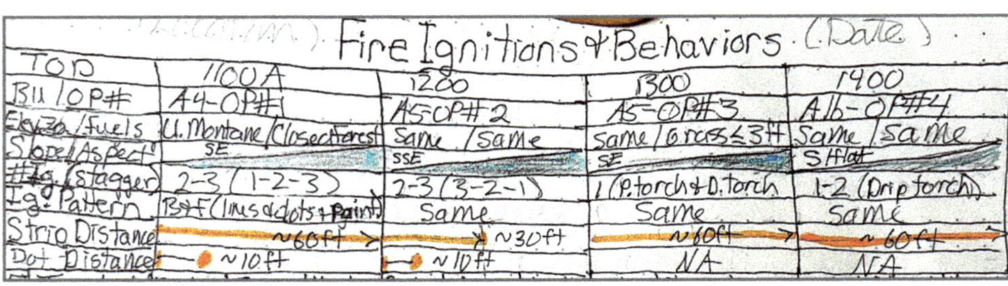

An hourly tracker organized specifically for ignition observations. Simple lines, symbols, and color-coding communicate changes in ignition tools, techniques, patterns, spacing, and crew organization, making relationships and patterns easier to recognize than tables of numbers alone.

Sketching Burn Activities

Sketching offers another way to document the human and ecological dimensions of good fire. While notes capture data, sketches capture interaction—who is present, what they are doing, and how fire moves through both people and place.

During another session at the Wild Wonder conference, I was inspired by Nishant Jain, who demonstrated quick, observational sketching techniques built around what he calls an *attention spiral.* His approach begins with a single point of interest and expands outward, adding context only as time and curiosity allow.

This method translates well to fire environments, where conditions are dynamic and time is limited.

Start with the essential action and only simple outlines:

- The curve of an arm and bend of a hand holding a drip torch or just the outline or silhouette of a person holding a drip torch
- The shape of flames coming off the recently ignited surface grass

Then expand outward:

- The texture and continuity of vegetation around the fire practitioner using the ignition tool
- The shape and direction of flame as it responds to wind and slope

From there, selectively add context:

- A few defining elements of the vegetation—grass stems, shrub structure, or litter layers
- Simple outlines of terrain or horizon
- Minimal washes of color to suggest heat, smoke, or light

Resist the urge to draw everything. Instead, focus on what tells the story most clearly. A well-placed line or a small patch of color can communicate more than a fully rendered scene. Over time, combining poetry and sketching builds a layered record—one that captures not only what fire *does*, but how it *feels* to be part of it.

Fire Behavior

Weather, terrain, fuels, and ignitions help us anticipate fire behavior. Observing fire behavior allows us to compare those expectations with the fire's actual response and refine our understanding for the next burn.

This is the moment when everything you've been observing begins to move. The weather you've felt on your skin, the fuels you've examined in your hands, the slopes you've climbed, and the ignitions you've watched being placed now come together as fire responds to the landscape. Flames lengthen and shorten, slow and accelerate, quietly back through shaded fuels or actively head across open grasses. These changing behaviors reveal the interactions among weather, terrain, fuels, and ignitions that until now have only been anticipated.

My focus is on what I call **"loosed" fire**—the stage when ignitions have taken hold and fire begins moving independently across the landscape. I include little discussion of extreme wildfire behavior and instead emphasize the fire behavior you are most likely to observe during prescribed and cultural burning.

Fire behavior observations provide an important connection between burn goals, objectives, burn parameters, and the fire's actual response. Measurements such as flame length, rate of spread, and other fire behavior characteristics help practitioners compare observed fire behavior with anticipated fire behavior, evaluate whether burn objectives are being met, and recognize when changing conditions may require operational adjustments. Just as importantly, these observations help refine future burn parameters, improve prescribed fire planning, and strengthen situational awareness through experience.

Throughout this chapter, you'll learn the common terminology used to describe fire behavior and the measurements used to observe it. These terms provide a shared language for communication, comparison, and evaluation that supports both operational decision-making and thoughtful journaling.

I Notice... how fire changes as it responds to weather, terrain, fuels, and ignitions, revealing relationships that were only anticipated before the burn.

I Wonder... why fire behaves differently across the burn unit and how those changes relate to burn goals, objectives, burn parameters, and operational decisions.

It Reminds Me Of... how movement, rhythm, and change reveal relationships that are difficult to see by observing weather, terrain, fuels, or ignitions alone.

Fire Behaviors

Fire behavior is often described by separating it into recognizable types and characteristics, but these are all parts of a single, connected fire responding to weather, terrain, fuels, and ignitions. Learning the common terminology helps you communicate with fire practitioners, compare observations, and understand how different parts of the fire contribute to the behavior of the burn as a whole.

The fire behavior terminology used in this section is adapted from the NWCG Glossary and emphasizes the fire behaviors you are most likely to observe during prescribed and cultural burning. The accompanying illustrations and journal examples are intended to help you recognize these behaviors in the field and relate them to burn goals, objectives, burn parameters, and the surrounding fire environment. More detailed discussions and illustrations can also be found in my earlier book, *Pyrosketchology: An Illustrated Guide to Nature Journaling About the Fire Environment*.

Throughout this section, I continue using my distinction between **"leashed"** and **"loosed"** fire. In the previous chapter, I used the terms **back, flank,** and **head** to describe intentional ignition placement. Here, I use **backing, flanking,** and **heading** to describe the fire's behavior after it has spread beyond the point of ignition. This is my own observational convention rather than standard fire terminology. I find it helps separate observations of ignition patterns from observations of fire behavior, making it easier to compare how ignition decisions influence the fire's response.

Observing Beyond the Terminology

Technical terminology gives us a common language. Your journal can also capture **how the fire behaves** and **what it seems to be doing**.

Backing fire → slow, gentle, lingering, patient

Flanking fire → steady, weaving, balanced, connecting

Heading fire → fast, energetic, advancing, cleansing

Spot fire → wandering, searching, unexpected

Flare-up → sudden, brief, pulsing

Rather than replacing the technical terms, these descriptive words can help you observe movement, relationships, and memorable moments that are difficult to capture with terminology alone.

The following fire behaviors are presented individually for clarity. During a burn, however, several of these behaviors often occur simultaneously in different parts of the burn unit.

Surface Fire Behavior

Surface fire behavior emphasizes the horizontal movement of fire across the burn unit as it responds to weather, terrain, fuels, and previous ignitions. Although fire is often described using individual behavior types, these are simply different expressions of the same fire. At any given moment, one portion of the burn may be quietly backing through shaded fuels while another heads across an open grassland, with flanking fire connecting the two. Together, these behaviors form the overall fire behavior observed across the landscape.

The fire behaviors presented in this chapter emphasize those most relevant to prescribed and cultural burning. I do not include many of the terms used to describe extreme wildfire behavior because the focus of this book is on observing, understanding, and journaling good fire. The examples that follow represent a range of low- to high-intensity fire behaviors that may be observed while working toward desired burn objectives.

As you observe, engage more than your eyes. Notice how flame length, rate of spread, sound, smoke, and heat vary across the burn unit. Compare these observations with the weather, fuels, terrain, burn parameters, and ignition patterns you've already recorded. In the next section, we'll shift from these primarily horizontal observations to crown fire behaviors, where fire movement is observed vertically in relation to ladder fuels and the forest canopy.

Smoldering

A flameless combustion phase that may occur when environmental conditions limit fire spread, or after the main flaming phase has passed and larger woody fuels continue to burn internally. Smoldering areas surrounded by unburned fuels can present a risk of reignition and should be noted in your journal.

Creeping Fire

Small, low-intensity flames moving without a clearly defined flame front. This behavior typically exhibits very slow rates of spread.

Backing Fire

An organized pattern of smaller flames moving against the wind or downslope. Backing fire generally produces lower intensity and slower spread, but longer heat residence time can increase impacts on root systems. Flame heights in prescribed burns typically range from 0 to 4 feet.

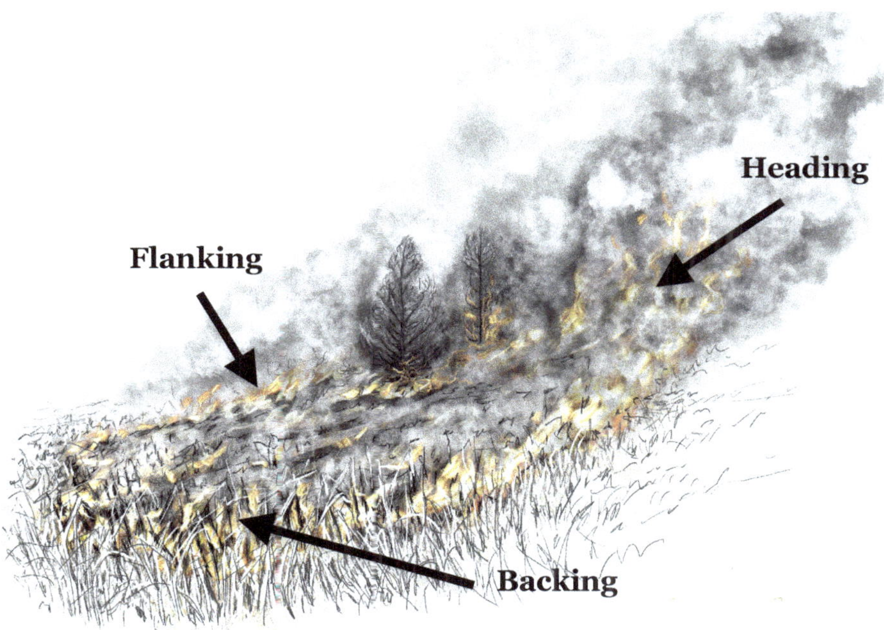

Flanking Fire

Fire moving parallel to wind direction or slope. This behavior often produces moderate intensity with variable flame heights and spread rates. Flame heights typically range from 0 to 15 feet in prescribed burns.

Heading Fire

Fire moving with the wind or upslope. This behavior is characterized by higher intensity, taller flames, and faster spread rates. Flame heights of up to 15 feet are common in prescribed burns.

Spot Fires

New ignitions outside the main fire area caused by wind-transported embers. hese most often originate from heading or flanking fire areas during higher-intensity burning.

Flare-Ups & Jackpots

Rapid increases in flame height and intensity in areas of concentrated fuels. These events may produce spot fires and should be recorded with time and location.

Fire Whirls

Rotating columns of gas, smoke, and flame that develop under unstable atmospheric conditions. Fire whirls vary in size and are sometimes confused with fire tornadoes, which are much larger and can generate localized weather effects.

Crown Fire

While surface fire behavior emphasizes the horizontal movement of fire across the burn unit, crown fire behavior focuses on the vertical movement of fire into the forest canopy. These behaviors develop as surface fire interacts with ladder fuels and overstory vegetation, creating important observations about vertical fuel arrangement, fire intensity, and the fire's response to changing conditions. Crown fire behavior may occur as individual trees torch, spread through patches of trees, or, under more extreme conditions, move independently through the canopy.

Observe not only whether crown fire occurs, but where it occurs. Compare these observations with your earlier notes about vertical fuel arrangement, ladder fuels, canopy characteristics, weather, terrain, and burn objectives. These relationships help explain why fire remained on the surface in some locations while moving into the canopy in others.

Torching / Passive Crown Fire
Occurs when individual trees or small groups ignite as surface fire moves through ladder fuels into the canopy.

Active Crown Fire
Occurs when multiple trees ignite and fire spreads through the canopy in conjunction with surface fire.

Independent Crown Fire
Unlikely during prescribed burns. This occurs when crown fire spreads independently of surface fire.

Fire Measurements

The fire behaviors described in the previous section help us recognize and describe how fire moves across the burn unit. Fire measurements provide a way to compare those observations with burn goals, objectives, burn parameters, and operational decisions. Rather than replacing descriptive observations, measurements help refine them by providing a common language for comparing fire behavior from one part of the burn unit to another and from one burn to the next.

Most fire behavior measurements used during prescribed and cultural burns are visual estimates rather than direct field measurements. Observers are rarely close enough—and should never move closer—to precisely measure flame length or rate of spread during an active burn. Instead, compare flames with nearby vegetation, the height of burn personnel, or other reference objects whose size is known or can be reasonably estimated. The body-based measurement techniques introduced in the Fuels chapter can also provide helpful reference points when estimating flame dimensions and spread.

Remember that these are estimates. Fire behavior is commonly evaluated within ranges of flame length, rate of spread, and fire intensity rather than as exact values. Estimating whether fire behavior falls within a desired range is often more useful than pursuing a precise measurement. Over time, comparing these estimates with burn objectives, burn parameters, and observed fire effects helps refine both your observations and your understanding of fire behavior.

General Fire Measurement by Fuel Loading

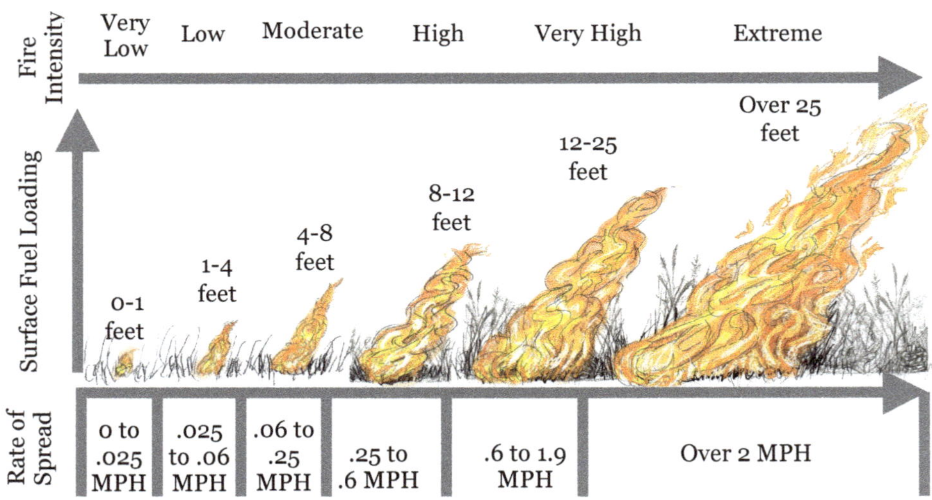

All flame/fire scenarios are based on conditions with 2/3rds of the surface fuels cured (grass, fallen leaves, and vegetation litter), little to no slope, and wind speeds 5 MPH or less.

Fire Intensity

Fire intensity provides a practical way to summarize and compare overall fire behavior. Rather than relying on a single measurement, intensity reflects the combined relationship among flame length, rate of spread, fuels, and heat release. For journaling purposes, estimating the general intensity category is often more useful than attempting precise calculations because burn objectives and operational decisions are commonly framed within ranges rather than exact values.

The generalized intensity chart on the following page illustrates how increasing fuel loads are commonly associated with increasing flame lengths and rate of spread under standardized weather and slope conditions. Your burn plan may define more specific fire behavior parameters for the fuels, weather, terrain, and objectives of that particular burn. Comparing your observations with those parameters provides a more meaningful assessment than relying on generalized values alone.

Flame Length & Flame Height

Fire management uses numerous fire behavior measurements, but for journaling purposes I focus on flame length and flame height. Although these terms are often used interchangeably in conversation, they are different measurements and may differ when wind or slope angles stretch the flame.

Flame Measurements

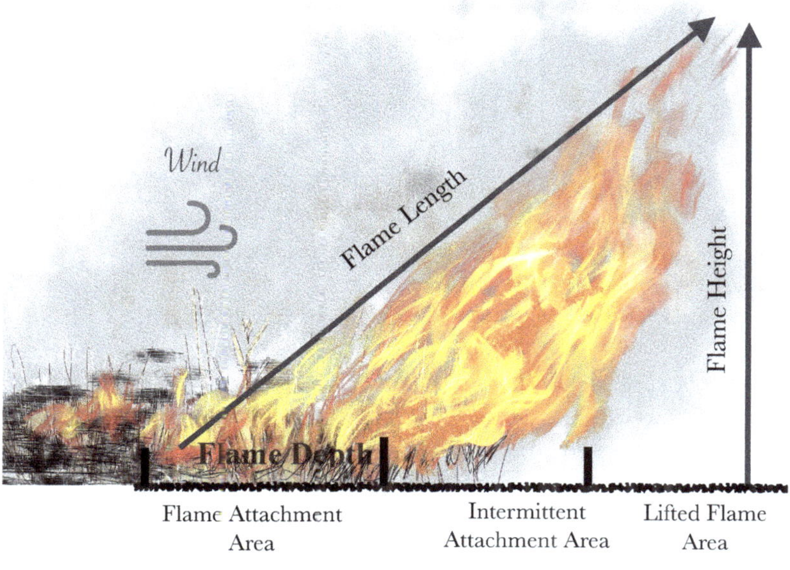

For journaling, flame length is especially useful for observing the horizontal movement of fire across the burn unit and considering how flames preheat and spread through surface fuels. Flame height is often more useful for observing the vertical movement of fire and its potential to influence shrubs, ladder fuels, and tree canopies.

I recommend estimating flame length and flame height visually. Comparing flames with nearby vegetation, the height of burn personnel, or other reference objects can improve the accuracy of your estimates.

Before recording flame measurements, compare your observations with the generalized fire intensity categories and the specific burn parameters for your burn. If the desired fire behavior is low intensity, for example, ask yourself whether the observed flame lengths appear to fall within that expected range. Use nearby vegetation, people, or other familiar reference objects to help estimate flame dimensions. You may choose to record both the estimated flame measurements and the corresponding fire intensity category, or record the intensity category when that level of detail is appropriate for your journaling purpose.

When practical, record separate flame length or flame height estimates for the backing, flanking, and heading fire areas at each Observation Point (OP). Comparing these measurements across different parts of the burn unit often provides more insight than a single estimate for the fire as a whole.

Rate of Spread

Rate of spread (ROS) describes how quickly the fire moves across the landscape. For journaling, I recommend estimating ROS by observing the different parts of the fire separately, particularly the backing, flanking, and heading fire areas. These comparisons provide more insight than estimating the movement of the fire as a whole because they help relate fire behavior to specific ignition patterns, burn parameters, and burn objectives. During a burn, these observations can help explain changes in ignition tactics, while comparisons made after the burn can improve future burn parameters, ignition decisions, and operational planning.

One simple approach is to identify two stationary reference points, such as trees, shrubs, rocks, or other landscape features, and estimate or measure the distance between them. Start timing when the flame front reaches the first reference point and stop when it reaches the second. For slower-moving backing fire, shorter distances—such as two measured sticks or two small rocks placed a known distance apart—often provide more practical observations than waiting for fire to travel a longer distance.

For faster-moving fire, rate of spread may be expressed in feet per minute or, for very rapid fire movement, miles per hour. For very slow-moving backing fire, inches per minute or inches per second may provide more meaningful observations. Select a measurement scale that best matches the speed of the fire and is practical for your observations. If your burn plan expresses desired fire behavior using different units, convert your estimates as needed so you can make meaningful comparisons with burn parameters and objectives.

When practical, record separate ROS estimates for the backing, flanking, and heading fire areas at each Observation Point (OP). Comparing these observations with flame length, fire intensity, burn parameters, and ignition patterns often provides a clearer understanding of how the fire is responding across the burn unit.

ROS Hack

Multiply the wind speed by 10% in forest and scrublands and by 20% in grasslands for the overall fire ROS (Cruz et al., 2022).

Fire Behavior Objectives & Parameters

Throughout this chapter, you've learned to observe fire behavior through terminology, measurements, and comparisons. Burn objectives and burn parameters provide the context that gives those observations meaning. Rather than simply recording flame lengths or rates of spread, compare your observations with the desired fire behavior identified for that burn. This comparison helps determine whether burn objectives are being achieved and whether operational adjustments may be needed.

Many burn plans specify desired ranges for flame length, rate of spread, and fire intensity, often by Fuel Model/Vegetation Type. These burn-specific parameters provide a more meaningful framework for observation than generalized fire behavior tables because they reflect the objectives, fuels, weather, terrain, and operational considerations of that particular burn. Generalized parameters remain useful when burn-specific information is unavailable, but they should be viewed as a broader point of reference rather than the desired outcome.

The examples that follow illustrate this difference. Notice how the burn-specific parameters define a much narrower range of desired fire behavior than the generalized examples. This greater precision helps guide observations, compare fire behavior with objectives, and recognize when changing conditions or ignition tactics may be needed.

Example Timber Litter Fuel Model Fire Behavior Parameters

Fire Intensities	Low	Moderate	High
Flame Length	<0.2 feet	0.3 to 0.5 feet	>0.6 feet
Rate of Spread	< 12 in/min	12 to 119 in/min	>120 in/min

Generalized Fire Behavior Parameters

Fire Intensities	Low	Moderate	High
Flame Length	<4 feet	4 to 8 feet	>8 feet
Rate of Spread	<.03 to .06 MPH	1 to 2 MPH	>2 to 6 MPH

Not all burn objectives are expressed through measurements alone. While prescribed fire plans often define desired flame lengths, rates of spread, and fire intensity, cultural burning may also describe desired fire behavior in more relational ways. Fire may be introduced at the right place, at the right time, and with the right tool, or described as moving gently through an area, cleansing the forest floor, or allowing fire to find its own path. These descriptions complement rather than replace fire behavior measurements. Together they provide additional ways to observe whether fire is behaving as intended and fulfilling the stewardship objectives of the burn.

Whether you are journaling through Story, Study, or Situational Awareness, burn objectives provide valuable context for your observations. Story observations help explain why fire behaved as it did and how people responded. Study observations compare measured fire behavior with objectives and parameters. Situational Awareness observations recognize when changing fire behavior approaches operational limits or requires increased attention. Together, these perspectives help transform measurements and observations into a more complete understanding of how the fire responded throughout the burn.

Reflecting on Fire Behavior Experienced

Throughout this chapter, you've learned to observe fire behavior through terminology, measurements, and comparisons. These approaches help describe what the fire is doing. Equally important is recognizing how you experience those same fire behaviors through your senses and how those experiences influence what you notice, record, and remember.

Fire behavior is not only observed with your eyes. It is heard in the crackle of grasses, the popping of shrubs, the hiss of moisture escaping from vegetation, and the growing roar of larger flames. It is felt in changing heat, air movement, and the need to step closer or farther away. As fire behavior changes, so does the information available through each of your senses.

Your own responses are observations as well. During lower-intensity fire, you may distinguish individual sounds, flame movements, and subtle changes in heat. As fire intensity increases, those details often blend together. Wind may mask individual fuel sounds, smoke may reduce visibility, heat may encourage you to reposition, and the growing intensity of the fire may narrow your attention. Recognizing these changes helps you understand both the fire and your ability to observe it.

Reflecting on fire behavior means observing both the fire and yourself as an observer. This awareness strengthens Story, Study, and Situational Awareness journaling by helping you recognize what you observed clearly, what became more difficult to distinguish, and how your observations may change as fire behavior evolves.

Fire Acoustics

Sound is often one of the first ways we recognize changing fire behavior. Long before we measure flame length or rate of spread, our ears begin gathering information about the fuels burning, the intensity of the fire, and changing wind conditions. Listening carefully can strengthen both situational awareness and journaling observations.

With experience, you may begin distinguishing the sounds of backing, flanking, and heading fire, as well as differences among Fuel Models/ Vegetation Types. Individual grasses, sticks, shrubs, and tree species often produce their own sound characteristics, while drought stress, fuel moisture, and fire intensity further influence what you hear. As fire grows in intensity or wind speeds increase, those individual sounds may gradually merge into a broader roar, making them more difficult to distinguish. Recognizing when this transition occurs is itself a valuable observation because it reflects both changing fire behavior and your ability to distinguish individual sounds.

As you listen, notice changes in:

- Fuel Model/Vegetation Type
- Fire behavior (backing, flanking, heading)
- Fire intensity
- Wind conditions
- Your ability to distinguish individual sounds

The following descriptive terms were used by **Kara M. Yedinak** in her fire acoustics research and may provide a helpful vocabulary for describing what you hear in your journal:

- Crackling (like sparkler fireworks)
- Pops or snaps (like a fireplace)
- Puff or whoosh *(like pulsing air)*
- Roar (like rushing water)

As you journal, consider recording:

- Descriptive words
- Sound intensity (volume)
- Rhythm or tempo
- Pitch or frequency
- Comparisons using prompts such as *It reminds me of...*
- Words, numbers, symbols, and simple sketches

The example on the next page was inspired by notes I recorded after attending a webinar presented by **Kara M. Yedinak** on fire acoustics. I expanded upon her research by exploring how those concepts could be observed and journaled in the field, integrating words, numbers, symbols, and simple sketches to describe, compare, and remember sounds that are often difficult to articulate through words alone.

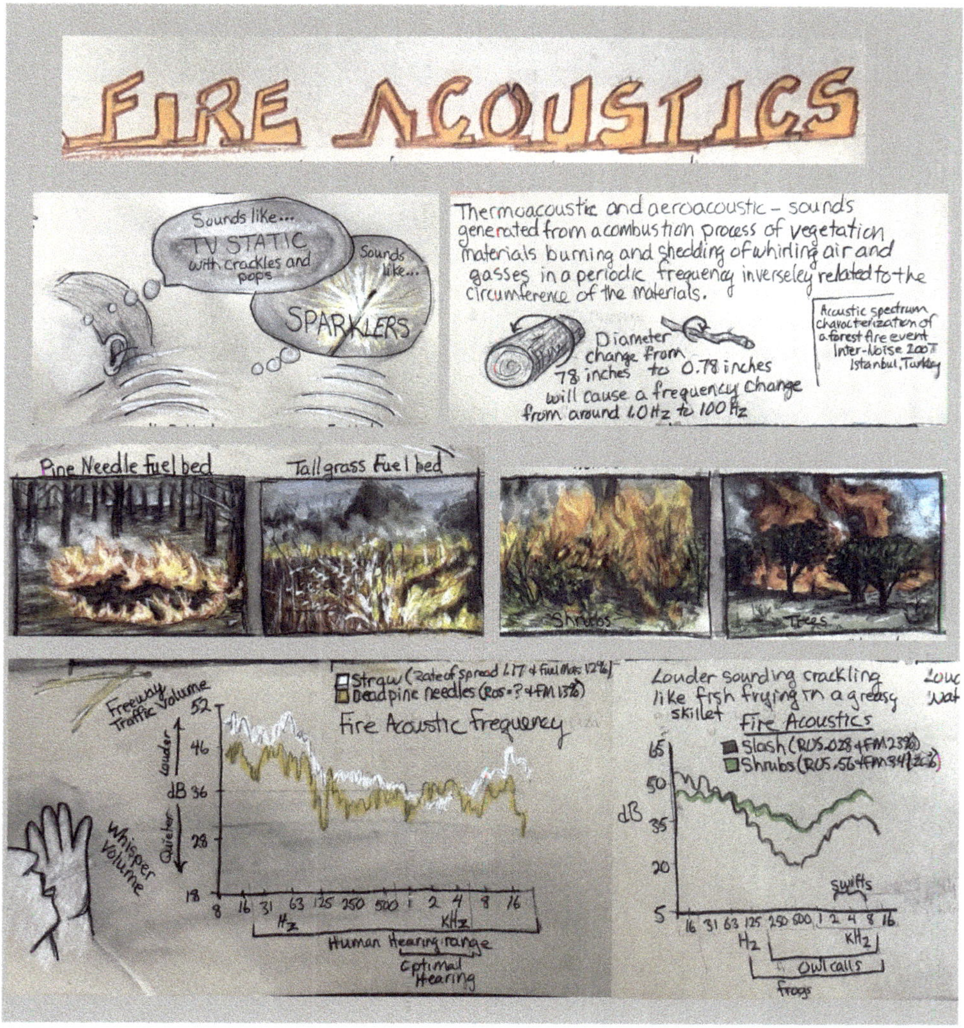

Journaling Fire Behavior

Fire behavior is often the point where all of your previous observations come together. Weather, terrain, fuels, ignition patterns, and burn objectives are no longer separate considerations—they become visible through the movement of fire across the burn unit. Your journal provides a place to organize those observations, compare them over time and across the landscape, and better understand how the fire responded to changing conditions.

No single journal format works for every burn or every purpose. Some journals emphasize the story of the burn, others focus on measurements and comparisons, while many combine Story, Study, and Situational Awareness observations. The examples that follow illustrate several approaches for organizing fire behavior observations using maps, trackers, sketches, photographs, and concise field notes. As with previous chapters, remember that journals are living documents. Returning to add notes, sketches, measurements, or reflections after the burn often strengthens both understanding and memory.

Example 1 – Hourly Fire Behavior Tracker

<u>Best suited for Study and Situational Awareness.</u>

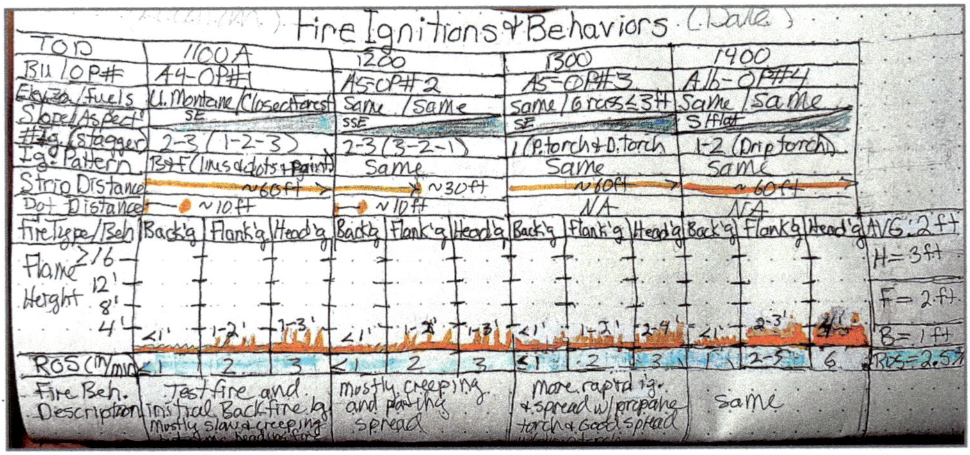

Hourly trackers help organize fire behavior observations into a format that makes changes through time easier to recognize. Rather than searching through pages of notes, patterns become visible as colors, symbols, and simple graphics reveal changing fire behavior throughout the burn. I often use color to visualize flame length ranges so increasing or decreasing fire intensity can be recognized at a glance.

My hourly tracker typically includes:

- Time
- Burn Unit / Observation Point (OP)
- Elevation Zone / Fuel Model
- Slope / Aspect
- Number of Igniters
- Ignition Pattern

- Strip or Dot Distance
- Fire Type / Behavior
- Flame Height / Length
- Rate of Spread

Journal size influences how much information can be recorded. Smaller journals increase mobility but limit the amount of data visualization possible. I often use color highlighting, symbols, and simple graphics to make patterns easier to recognize and compare over time.

Examples 2 & 3– Comparing Fire Behavior

Best suited for Story and Study. See examples on the next page.

Comparing observations from different locations or times often reveals relationships that may otherwise be overlooked. The following journal examples compare fire behavior in shaded and sun-exposed areas, as well as differences between low-intensity fire moving through green grasses and flowers and more active fire behavior within a narrow drainage. Together, these comparisons help explain how weather, terrain, fuels, and fire behavior interact across the burn unit.

Mapping Fire Behavior Idea

Best suited for Study and Situational Awareness. No visual

Fire behavior can also be organized spatially using the burn unit map. Rather than mapping ignition patterns, consider mapping observed fire behavior. A simple burn unit outline, vellum overlay, or transparent sticky note can be used to record areas of differing fire intensity. For example, yellow may represent lower-intensity fire behavior, orange fire behavior within desired burn parameters, and red areas of higher-intensity fire behavior. Comparing these patterns with terrain, Fuel Models/Vegetation Types, ignition patterns, and burn objectives often reveals relationships that are difficult to recognize from written notes alone.

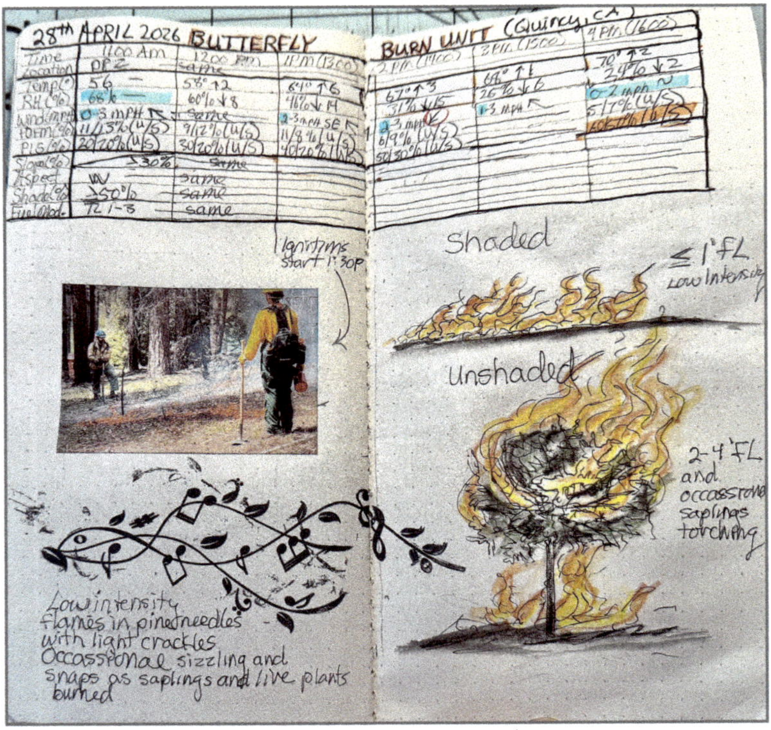
28th APRIL 2026 BUTTERFLY BURN UNIT (Quincy, CA)
Ignitions start 1:30P
Shaded
≤ 1' FL
Low intensity
Unshaded
2-4' FL and occassional saplings torching
Low intensity flames in pineneedles with light crackles. Occassional sizzling and snaps as saplings and live plants burned

Good lift during most of the burn
1457 hrs Ignitions in drainage
Some flames moving up trees
15:45 hrs
Scarlet Fritillary in the burn lower on the slope with patches of green grass.
Burn unit previously masticated and 10 hour plus larger dead & down was scattered thru the unit.

Reflecting Through Fire Behavior

Fire behavior is often described through measurements, categories, and terminology. These observations help us compare burn objectives, refine future prescriptions, and improve stewardship. Yet beyond the measurements lies another way of seeing. Fire behavior expresses the relationships among weather, terrain, fuels, ignitions, and people through the movement of heat and energy across the landscape.

Throughout this chapter, you estimated fire behavior from weather, terrain, fuels, and ignition patterns, then compared those predictions with what actually occurred. Through measurements, sensory observations, and journaling, each burn becomes an opportunity to observe, compare, and refine your understanding. Those observations not only help evaluate whether burn objectives and parameters are being achieved, but also strengthen future observations, improve burn prescriptions, and build the experience needed to recognize subtle changes in fire behavior before they become significant.

The Ecological Dance

An ecosystem is never still. Elements cycles through the atmosphere and the earth. Whether we notice it or not, the Earth system is always changing—always dancing.

Within this ongoing dance, every partner responds to the others. Weather shapes the rhythm, terrain guides the movement, fuels carry the memory of seasons past, and people influence the choreography through their relationships with fire. Fire does not begin the dance; it joins it as a recurring partner in one of nature's oldest and most transformative duets.

While fire reveals the transformative movement of a landscape, smoke reveals what came before and hints at the future. It rises with the fire, drifts with the weather, settles with the landscape, and carries another chapter of the ecological story waiting to be observed. Before we explore the changes fire leaves on the landscape, let us first follow what it releases into the sky.

Smoke

Fire transforms the living dance of the landscape. Smoke carries the visible exhalation of that transformation into the atmosphere. Its colors, shapes, movements, and scents reveal clues about the fuels being consumed, the fire behaviors unfolding, the weather and atmospheric conditions, and the effectiveness of fire operations.

Smoke rises with the fire, drifts with the weather, settles with the landscape, and continues interacting with the atmosphere long after the flames have passed. In this chapter, I invite you to lift your gaze and follow that journey. Rather than observing smoke as simply a byproduct of fire, learn to see it as another expression of the relationships among fire, fuels, weather, terrain, and people.

Throughout this chapter, we will use spatial and temporal journaling techniques to observe and interpret smoke across the landscape. Much of the foundational information is drawn from the *NWCG Smoke Management Guide* and related training materials, while the journaling approaches emphasize observation, comparison, and the relationships between smoke, fire behavior, and the atmosphere.

I Wonder...
What might smoke reveal that I cannot learn from watching the flames alone?

Smoke Color

Smoke color is often the easiest smoke characteristic to notice, whether you are standing beside the fire or observing from miles away. Rather than memorizing colors, begin by asking what the smoke might be telling you about the fuels, combustion, weather, stage of the burn, or even fire operations. For example, smoke often becomes noticeably whiter when water is applied during holding activities as water vapor mixes with the smoke.

Before focusing on individual smoke colors, observe how smoke color varies across the landscape and throughout the burn period. Different fire behavior areas, **fuel models (vegetation types),** and fire operations often produce a mosaic of smoke colors and concentrations within the same burn unit. As fire moves across the landscape and conditions change through the day, those patterns may also change. Look for patterns rather than isolated colors. Ask yourself where lighter and darker smoke occurs, how those patterns change through time, and what those changes might reveal about the relationships among fuels, fire behavior, weather, terrain, and fire operations.

I recommend focusing first on **color saturation** or **consistency**—how light or dark the smoke appears—before trying to identify specific colors. In general, darker smoke often indicates larger or more concentrated smoke particles.

White Smoke

White smoke commonly indicates efficient combustion in fine fuels such as dry grasses, but it may also represent water vapor. It is often observed at the leading edge of a fire where vegetation is heated and dried before ignition, in smoldering areas where combustion is largely flameless, or during holding operations when water is applied.

White coloration is also commonly visible at the tops of towering smoke columns, where water vapor mixes with heated smoke and may contribute to pyrocumulus cloud formation.

Key observations:

- **White smoke near the ground** — Often associated with moisture from surface fuels, smoldering combustion, or water application during holding.
- **White smoke high in the plume** — Often associated with atmospheric moisture mixing into the smoke column.

Gray or Brown Smoke

Moderately saturated gray or brown smoke often indicates the consumption of heavier fuels, producing greater concentrations of carbon particles than fine grass fuels. Variations in gray or brown coloration may also reflect changing combustion efficiency, fuel moisture, or transitions between fuel models and vegetation types.

Dark Brown or Black Smoke

Highly saturated dark brown or black smoke often indicates dense fuel concentrations or fuels containing waxes, oils, or other compounds that produce greater concentrations of soot and carbon particles. Dark smoke may also result from burning structures, vehicles, or other human-made materials. Remember that sun angle and shadows can also make portions of a plume appear darker than they actually are.

Smoke Concentration

Different fire behavior areas often produce different smoke concentrations. Compare smoke produced by backing, flanking, heading, and smoldering fire areas, and consider how these differences may influence visibility, roads, nearby communities, or holding operations.

General smoke concentration categories:

- **Light** – Often white; common in backing fire and smoldering areas.
- **Moderate** – Often white, gray, or brown; common in flanking fire areas.
- **Heavy** – Often dark gray or black; common in heading fire areas and concentrated fuel pockets.

Smoke Locations & Movements

Much of smoke observation centers on shape and movement, but these patterns can be more challenging to recognize when fire moves across a landscape rather than rising from a single point source. Many plume diagrams were originally developed from observations of point sources, such as chimneys or industrial stacks. In **wildland fire**, however, smoke often originates from multiple ignition areas and fire behavior zones spread across varied terrain.

Rather than looking for a perfect plume type, observe where the plume displays characteristics of one or more categories. Heading fire areas often produce the strongest vertical plume development, while backing, flanking, and smoldering fire areas may contribute lighter or lower smoke that blends into the overall plume. Terrain, wind, and changing fire behavior continually reshape these patterns as the burn progresses.

Think of plume types as tools for interpreting atmospheric conditions rather than exact labels. As you observe smoke across the burn unit, compare where it lifts, spreads, settles, or becomes trapped. These relationships often reveal more about smoke behavior and atmospheric stability than identifying a single plume category.

Smoke Plume Types

Smoke plumes can be categorized by height, shape, moisture content, and movement which is based on atmospheric conditions. Many visual examples resemble point-source plumes, so adapt these concepts to broader prescribed-fire patterns, particularly observing the heading-fire plume characteristics.

Fumigating Plume

Typically rising to around 2,000–3,000 feet, this plume gives the impression of lifting at first, then slowly collapsing back toward the ground. From a distance, it appears as a diffuse, sagging layer—like a ceiling lowering unevenly over the landscape. The smoke loses its vertical structure and spreads outward in a soft, downward drift, often filling low areas. Near the surface, it can feel heavy and enveloping, with visibility decreasing as the smoke settles and lingers.

The following plume types are visualized in a combined chart further along in this section.

Fanning Plume

This plume stretches outward in a thin, flattened sheet, often between 1,000 and 5,000 feet. It resembles a narrow ribbon or the blade of a fan extending downwind. The edges are relatively smooth and defined, with very little vertical motion. Instead of rising or falling, the smoke slides laterally across the sky, maintaining a consistent height. From the ground, it can appear as a long, horizontal band that seems to skim the atmosphere, spreading far while staying tightly compressed.

Coning Plume

A coning plume forms a balanced, symmetrical shape, typically extending from 5,000 to 10,000 feet. It rises steadily and then expands outward in all directions, creating a triangular or cone-like form that widens with distance. The edges are soft but structured, and the smoke appears evenly distributed. Movement is smooth and continuous, with no abrupt shifts—both upward lift and outward spread feel steady and proportional.

Looping Plume

This plume is highly dynamic, usually between 3,000 and 6,000 feet, with smoke that rises and falls in visible waves. It looks like a series of rolling curls or looping arcs—almost like a slow-motion roller coaster. Dense gray puffs surge upward, then dip back down, creating alternating bands of thicker smoke and clearer air. The motion is irregular and energetic, with constant vertical oscillation that gives the plume a restless, pulsing appearance.

Lofting Plume

A lofting plume rises cleanly and confidently, often reaching heights of 3,000 to 15,000 feet. The smoke column lifts upward with minimal downward drift, maintaining a strong vertical structure before bending gently with upper-level winds. It can appear tall and well-defined near the source, then gradually fan out at higher elevations. From below, the air near the ground often remains relatively clear, while the bulk of the smoke is carried high above, moving efficiently across the landscape.

Trapping (Layered Plume)

In a trapping scenario, smoke becomes confined within a horizontal layer, typically between 2,000 and 8,000 feet. Visually, it appears as a dense, flat slab or band stretching across the sky. The top and bottom boundaries are often well-defined, giving the layer a compressed, stacked look. Movement is almost entirely lateral, with the smoke spreading outward but showing little to no vertical development. Over time, this layer can thicken and persist, creating a heavy, persistent layer.

Dry Convection Column

A convection column occurs when fire intensity generates a strong convective heat column, with the smoke plume held within the column. Smoke and heat columns can reach very far into the atmosphere with lofting and coning shapes. A dry convection column occurs without sufficient moisture to form cloud structures. These columns may extend into mid- or high-level cloud zones.

Moist Convection Column

Similar to dry convection columns but with higher moisture levels from fuels or the atmosphere creating billowing clouds at the top or separated from the smoke column. Represents an early stage of pyrocumulus development.

Pyrocumulonimbus Convection Column

Occurs under highly unstable conditions with significant moisture. These smoke columns can generate localized weather effects, including thunderstorms and strong downdraft winds. Picture a giant smoke column with a thunder cloud on top.

These visual patterns become easier to recognize with practice. By focusing on shape, motion, and how the plume occupies space, you can quickly interpret how smoke is behaving in real time—often before instruments or forecasts confirm it.

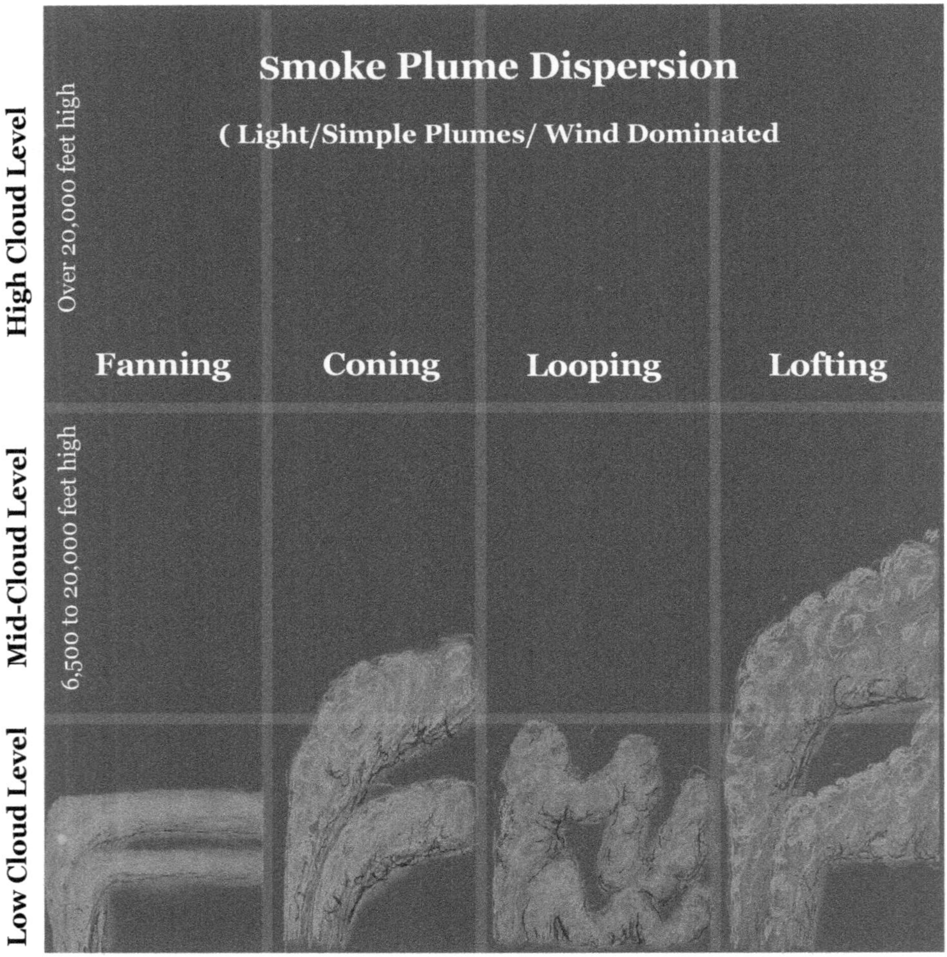

Smoke Plume Cycles & Relationships

Smoke behavior is closely tied to the daily rhythm of weather and atmospheric stability. Temperature, relative humidity, wind speed, cloud cover, and solar heating all influence how smoke forms, rises, and moves across the landscape. Burn plans may include weather parameters such as temperatures above 70°F, relative humidity below 60%, wind speeds below 7 MPH, and less than 50% cloud cover (NWCG PMS 477). These conditions influence not only ignition and fire behavior, but also smoke development, plume shape, and dispersion.

Throughout the day, changing atmospheric conditions create recognizable patterns in smoke behavior. During the cooler, more stable conditions of early morning and evening, smoke tends to remain lower to the ground, moving laterally and collecting in drainages and other low-lying areas. As solar heating increases and the atmosphere becomes more unstable, smoke lifts, mixes, and develops stronger vertical movement. By late afternoon, plume development often reaches its greatest height and dispersion before gradually lowering again as temperatures cool and atmospheric stability returns.

These daily cycles are closely reflected in smoke plume characteristics. Well-developed vertical plumes are generally associated with greater atmospheric instability and stronger mixing, while flatter or layered plumes often indicate more stable conditions and limited vertical development. As smoke rises, upper-level winds increasingly influence plume direction, tilt, and transport, while cloud cover can either enhance or limit surface heating and plume development.

As you observe smoke throughout the burn, compare changes in plume type, height, color, movement, and mixing from one observation period to the next. Notice when smoke begins lifting, when it starts settling, or when plume characteristics shift with changing weather conditions. These observations connect the fire on the ground with the atmosphere above, revealing daily patterns that repeat while remaining uniquely shaped by weather, terrain, fuels, and fire behavior.

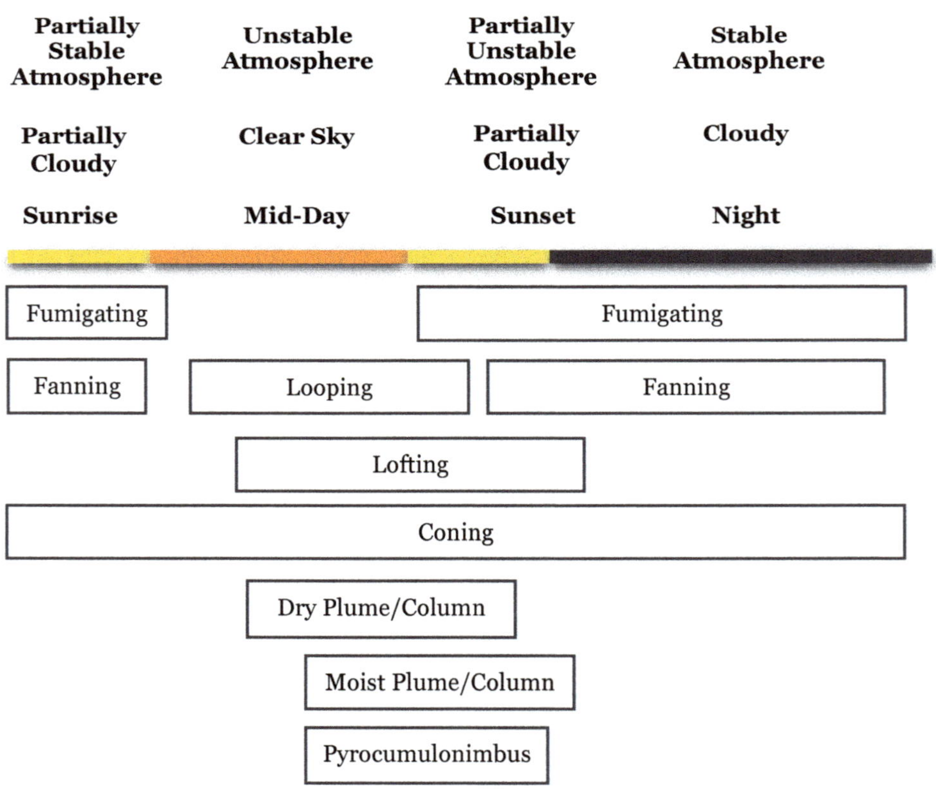

Smoke Mixing/Dispersion Quality

While smoke plume types describe the shape and movement of smoke, **smoke mixing and dispersion describe how effectively the atmosphere lifts and distributes that smoke away from the ground.** Although the two observations are closely related, they are not the same. Similar plume types may occur under different mixing conditions, making it helpful to consider plume characteristics together with atmospheric stability and smoke dispersion.

Higher smoke columns are often associated with more intense burning and greater total smoke production, but they also generally indicate better vertical mixing and fewer near-surface air-quality impacts. In prescribed and cultural burning, fire intensity is often lower, producing less smoke overall but sometimes allowing smoke to linger closer to the ground. Burn plans typically include smoke management strategies to balance burn objectives with air-quality concerns.

Mixing height goals commonly target approximately **1,500–1,700 feet** to support effective dispersion (NWCG PMS 420-2).

Smoke Mixing Categories (Simplified for Journaling)

Excellent / Good (Category 6)
Unstable atmospheric conditions with strong vertical lift.

- Dry or moist convection columns

Fair / Moderate (Categories 4–5)
Partially stable conditions with moderate lift.

- Daytime lofting plumes
- Some daytime fanning or coning plumes

Poor / Moderately Poor (Categories 2–3)
Stable conditions with minimal vertical mixing.

- Looping plumes with limited lift
- Nighttime lofting plumes
- Coning plumes with low fire intensity
- Column collapse events

Very Poor (Category 1)
Highly stable conditions.

- Fumigating plumes

Smoke Journaling

Smoke observations become more meaningful when they are compared across **time**, **place**, and **weather conditions**. Rather than trying to document every detail, choose a journaling approach that helps you recognize patterns in smoke behavior and the relationships among fire, weather, terrain, and fuels. The examples that follow illustrate several ways to organize those observations.

Examples 1 & 2. Hourly Smoke Tracker

Because journal space is limited, simplified plume shapes and height indicators can effectively represent smoke patterns without requiring detailed drawings. If you choose to include simple smoke sketches, add only a few essential details that capture the plume's overall shape and movement.

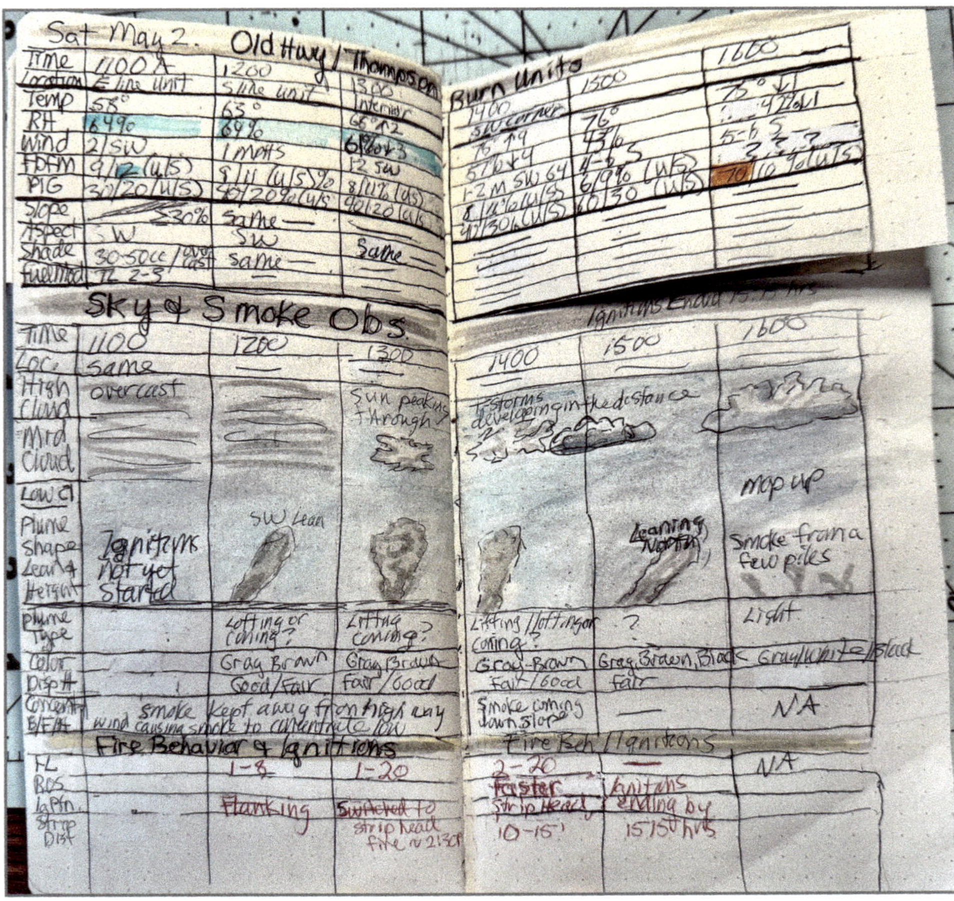

The key to integrating smoke observations into an hourly tracker is deciding which related data you want to combine. You may choose to create a dedicated smoke tracker or integrate smoke observations with broader weather and sky observations, depending on your journaling goals.

To provide context for smoke plume symbols or sketches, use relative cloud-layer heights marked on the tracker and/or nearby terrain features to estimate plume height. These visual references make it easier to compare plume development, movement, and atmospheric mixing throughout the burn period.

Include:

- Time and OP identifiers
- Smoke plume height relative to cloud levels
- Plume type and mixing/ dispersion category
- Smoke concentration by fire behavior zone
- Notes on smoke impacts

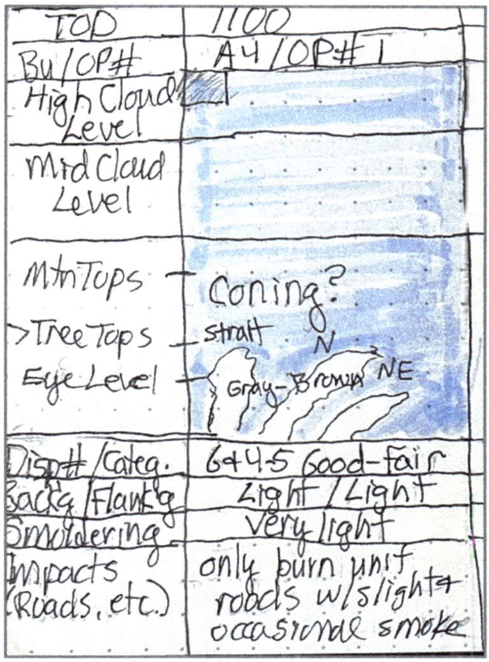

Smoke color may be recorded with a small annotation or directly on the sketch. Wind direction or plume lean can also help explain changing plume movement and atmospheric conditions.

The goal is not to create a realistic drawing, but to capture the dominant plume characteristics and compare how they change from one observation period to the next.

Example 3. Smoke Mapping

If time permits, mapping smoke across the landscape can reveal relationships that are difficult to recognize from a single observation point. Transparent overlays placed over burn unit maps or landscape sketches provide a simple way to compare smoke movement with terrain, fuels, and fire behavior. See my journal example on the next page.

I like using white acrylic pens with black ink on transparent overlay paper so smoke sketches remain visible against the underlying map without obscuring topography, fuel models (vegetation types), or control lines.

These sketches are not intended to show exact plume height or create realistic illustrations. Instead, they capture movement, shape, and behavior—how smoke lifts, spreads, settles, or oscillates across the landscape. Simple lines, arcs, shaded bands, and directional arrows are often enough to communicate plume movement and wind influence.

Comparing overlays from different burn periods or burn units helps reveal recurring patterns and strengthens your visual memory of smoke behavior.

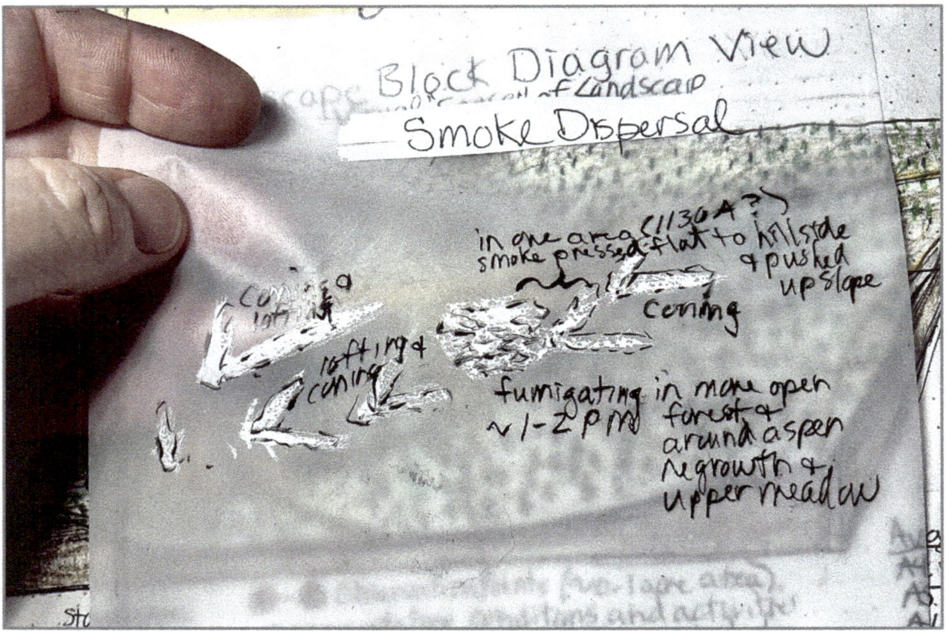

Example 4: Air Quality Journaling

Air-quality observations add valuable context. Visual range estimation provides a practical field method, though it can be influenced by humidity, contrast, and particle size (Oregon DEQ, 2023).

Air Quality Observation Tips

- Establish baseline visibility under clear conditions
- Face away from the sun when estimating visual range

- Use high-contrast background features
- Follow safety protocols during poor air quality

In my example, I used a subjective assessment and map references to estimate the distance I could see clearly. I then sketched the visible landscape features. After completing the sketch, I used a combination of public health references to estimate the air quality for that day based on how far I could clearly see objects.

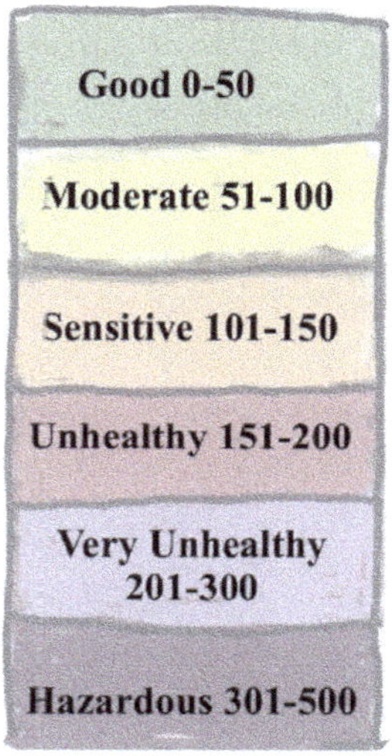

Good — Visibility up to 10 miles

Moderate — 6–9 miles

Unhealthy for Sensitive Groups — 3–5 miles

Unhealthy — 1.5–2.75 miles

Very Unhealthy — 1–1.25 miles

Hazardous — 0.75 miles or les

The journal page above highlights smoke science and its benefits to understory plant communities. While often viewed as a negative byproduct, smoke has important ecological and cultural roles. Consider journaling additional perspectives and insights on its value or negative impacts.

Fire Effects

Fire effects are the continuing story of the ecological dance. The flames have passed, yet transformation continues to unfold across minutes, days, seasons, and years. Some changes are immediate and easy to recognize, while others emerge slowly as soils, plants, wildlife, weather, and people respond to the renewed landscape.

This chapter invites you to observe not only what changed, but why those changes matter. Every burn begins with intentions—reducing fuels, improving wildlife habitat, restoring cultural resources, protecting communities, or renewing ecological processes. Fire effects help us understand whether those intentions are being realized and what new relationships are beginning to emerge.

As the landscape grows quieter, a different kind of observation begins. Bird activity shifts, smoke settles, sunlight reaches the forest floor, and subtle changes in sound, scent, and texture become easier to notice. Your journal becomes more than a record of burned vegetation or consumed fuels. It becomes a record of transformation through time, helping reveal how fire continues to shape the relationships among people, place, and the living communities that share them.

Burn Goals, Objectives & Ecological Indicators

Fire effects are best understood in relation to the purpose of the burn. Burn goals and objectives provide the context for interpreting what changed, why those changes matter, and whether the fire accomplished what was intended. Some objectives focus on measurable outcomes, such as reducing hazardous fuels, limiting tree mortality, or improving wildlife habitat. Others emphasize broader ecological relationships, cultural values, stewardship, or preparing the landscape for future care. Your observations become most meaningful when viewed through the lens of those intended outcomes.

Many fire effects can be observed immediately after the flames pass. Changes in fuel structure, vegetation, scorch, consumption, char, and soil conditions provide important information about how fire moved through the landscape. These structural observations form the foundation of most fire-effects monitoring and remain valuable whether you return hours, weeks, or months after the burn.

Ecological observations tell another part of the story. Wildlife, insects, fungi, flowering plants, and other living communities respond not only to changes created by fire, but also to the presence of people, smoke, heat, changing habitat, and the passing seasons. Some responses begin before ignition, others occur during active burning, and many continue developing long after the fire is out. Together, structural and ecological observations provide a richer understanding of how fire influences the living landscape.

I Wonder...

What relationships are already unfolding before ignition, and how might they change as fire moves through this landscape?

Bird Indicators

Birds are among the most useful ecological indicators because they respond quickly to changing conditions before, during, and after a burn. Their locations and behaviors help interpret vegetation structure, habitat relationships, seasonal timing, and the scale and vulnerability of fire effects. Begin observing birds as soon as you arrive at your Observation Point (OP), rather than waiting until the burn is complete.

Observe birds across both the horizontal landscape and the vertical arrangement of vegetation. Notice where they sing, forage, perch, or nest within the overstory, mid-story, understory, snags, shrubs, grasses, or surface fuels. These observations reinforce the vertical vegetation structure and fuel models (vegetation types) you have already documented while establishing a baseline for comparing post-fire habitat conditions.

Listening is often more effective than looking. Pause quietly for several minutes before recording observations. Notice songs, contact calls, alarm calls, wing sounds, movement between vegetation layers, and repeated use of perches or nest sites. As fire behavior and smoke conditions change, observe whether birds alter their behavior, relocate, or return. These responses become especially meaningful during sensitive periods such as territory establishment, nesting, fledgling development, and migration, when bird behavior may reveal both the immediate effects of fire and whether the timing of the burn coincides with periods of greater ecological vulnerability.

Incorporate bird observations into your fuel notes, Observation Point sketches, or burn unit maps. A few symbols, arrows, perch locations, or song descriptions are often enough to reveal meaningful relationships when compared before, during, and after the burn.

Birds are only one example of how seasonality shapes our understanding of fire effects. Flowering plants, fungi, insects, amphibians, reptiles, mammals, and many other organisms follow seasonal cycles that influence when they can be observed and how they respond to changing habitat conditions. Their presence, absence, life stage, and behavior often reveal ecological relationships that cannot be understood from vegetation observations alone.

Unlike structural fire effects, which can usually be observed immediately after a burn, many ecological indicators appear only during particular seasons or life stages. Returning to the same Observation Point (OP) after rainfall, spring green-up, flowering, insect emergence, or migration often reveals changes that would remain invisible on burn day. These observations also provide valuable context for burn timing, reminding us that similar fire behavior can produce different ecological outcomes depending on when the burn occurs within the annual cycle.

Because these indicators are seasonal, they are most useful when they are naturally active. The chart is intended as an observational guide rather than a calendar of fixed events. Use this in combination with the fuel model microclimate indicators to deeper your ecological perspective.

Bird Behavior Indicators

Observation	Indication
Repeated territorial singing  Stays close to same spot heard	Territories established; nesting season begins
 Birds carrying nest materials	Pair bonding and nest establishment
 Alarm calls, agitated body language, dive-bombing, injury-faking	Potentially incubating eggs, nestlings or fledglings close
Birds flushed/leaving nest 	Nestlings present; active brood rearing
 Birds carrying food away and not immediately eating	Nestlings present; active brood rearing
 Young birds in or near nest	Fledgling stage; young highly vulnerable
 Birds missing some feathers; feathers on the ground	Juvenile and adult molting; birds energetically stressed, flight may be limited

First-Order Fire Effects

Fire effects are commonly described by when they occur. **First-order fire effects** are the immediate consequences of fire on ecosystem elements during combustion and within hours afterward. They include changes such as vegetation scorch, fuel consumption, soil heating, ash deposition, and other direct effects of heat and flame.

Many fire professionals use first-order fire effects to evaluate whether burn objectives have been achieved. As you journal these observations, relate them not only to the fire behavior you observed, but also to the burn goals and objectives established before ignition. Fire effects become most meaningful when interpreted within that broader context rather than as isolated observations.

While field protocols often emphasize measurable changes, your journal can also include sensory observations, ecological indicators, and comparisons with pre-burn conditions. Together these observations create a richer understanding of how fire transformed the landscape.

Burn Severity

Burn severity is a qualitative description of the degree to which fire has altered vegetation, surface fuels, and soils. Although originally developed primarily for wildfire assessment, burn severity also provides a useful framework for describing fire effects following prescribed and cultural burns. It should not, however, be confused with achieving burn objectives. Similar burn severity may represent excellent outcomes for one burn and undesirable outcomes for another, depending on the intended purpose of the fire.

Several field references exist, including the *U.S. Forest Service Field Guide for Mapping Post-Fire Soil Severity* (2010), the BAER Soil Burn Severity Field Guide, and the *National Park Service Thematic Burn Severity Mosaic* (1991). These references commonly evaluate observations such as canopy scorch, tree mortality, surface fuel consumption, and soil burn severity.

For journaling, we adapt these same observational concepts while emphasizing comparisons between pre-burn and post-burn vegetation structure, fuel conditions, and burn objectives. The following observations are organized using the same vertical vegetation structure introduced in the Fuels chapter to support consistent comparisons before and after the burn.

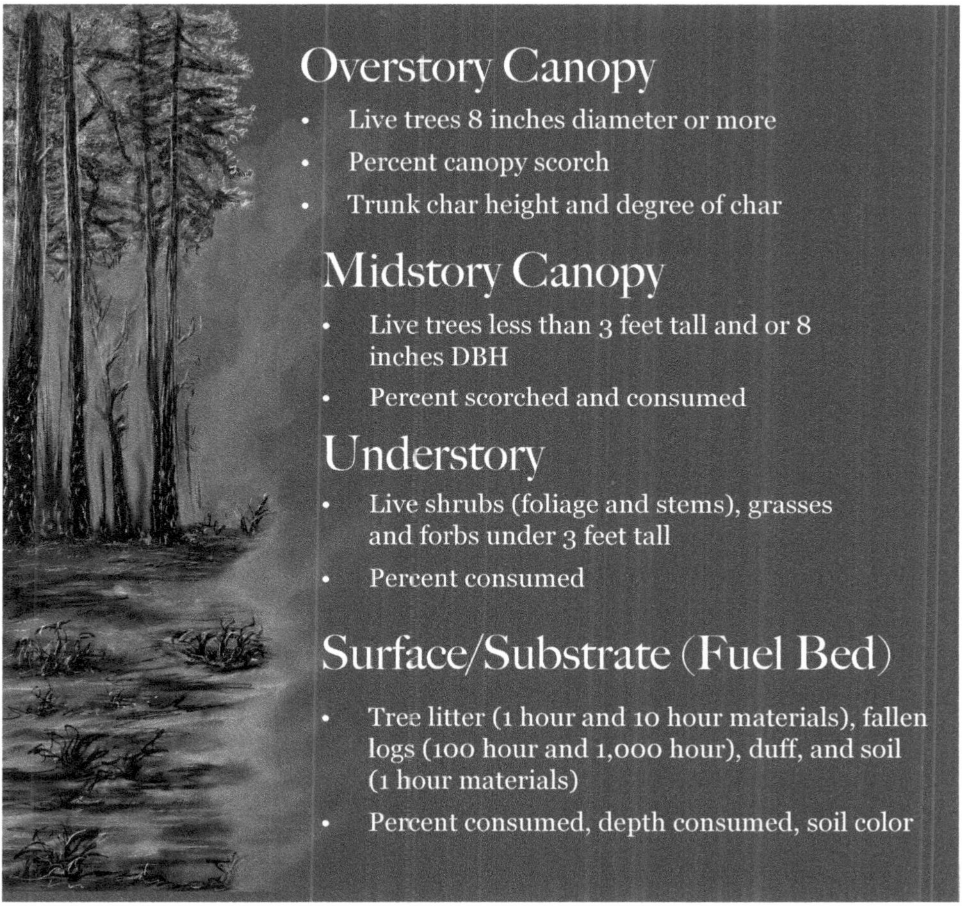

Vegetation/Fuel Burn Severity

Vegetation and fuel burn severity describe how fire has altered the structure and condition of vegetation from the overstory canopy to the soil surface. Rather than relying on a single observation, compare changes throughout the vertical arrangement of vegetation using the same structural layers introduced in the Fuels chapter. This approach helps relate fire effects to burn objectives while creating a consistent framework for before-and-after observations.

If you are observing only on burn day, many assessments will rely on the remaining vegetation and fuels. Compare remnant materials with your pre-burn observations whenever possible, or estimate pre-burn conditions using nearby unburned areas with similar vegetation. Consistent observations are often more valuable than highly precise measurements, especially when journaling rather than conducting formal monitoring.

The visual guide that follows adapts concepts from BAER and National Park Service field assessments while simplifying them for field journaling. These observations are framed most closely around the **Study** journaling approach, where consistent comparisons and repeated observations help document fire effects over time. If you are using a **Story** approach, you may choose to record only the observations that best support your narrative rather than completing each severity category. Each vertical vegetation layer emphasizes the fire effects most appropriate for that part of the fuel profile. For example, canopy observations focus on scorch and char, while shrub and surface observations emphasize consumption. Surface observations also combine estimates of the area affected with changes in litter and duff depth to better represent how fire altered the fuel bed.

Use the same vertical categories throughout your journal to compare pre-burn fuels, observed fire behavior, burn severity, and post-fire effects. Whenever possible, make these observations from a consistent **Observation Point (OP)** so the same area can be compared over time. If each OP is associated with a dominant fuel model (vegetation type), you can estimate average fire effects for that fuel model and compare how different vegetation communities responded to the burn. Observations from multiple OPs can then be combined to estimate overall burn severity for the burn unit. Each vertical category should include a **No Effect** option alongside **Low**, **Moderate**, and **High** severity to support consistent comparisons.

The overstory often shows the least immediate consumption but provides important clues about fire intensity, tree stress, and potential long-term survival. Using the illustrations and comparison table, estimate the overall fire effects for the overstory within your Observation Point (OP) rather than focusing on individual trees. Look for recurring patterns across the canopy and mature tree trunks, recognizing that fire effects may vary among trees even within the same OP.

Overstory

Canopy Scorch Severity

Low Severity	Moderate Severity	High Severity
<20% of Canopy	20-80% of Canopy	80-100% of Canopy

Trunk Char Height Severity

Low Severity	Moderate Severity	High Severity
<5 feet Char	5-9 feet Char	10 feet or higher Char

Bark Degree of Char Severity

Light Char	Medium Char	Heavy Char
Spotty char or scorch with scattered pitting of bark	Continuous charring with areas of minor reduction in bark thickness	Continuous charring, pronounced reduction in bark thickness with underlying wood sometimes exposed

Overstory Observations

Estimate the following observations within your OP:

- Percent canopy scorch on trees greater than 8 inches DBH
- Average number (or proportion) of scorched overstory trees
- Average canopy scorch percentage. Tree mortality may sometimes be estimated from crown scorch, although responses vary among species. Many trees experience significant stress, and some species have increased mortality when crown scorch exceeds approximately 75%.
- Average trunk char height
- Average degree of bark char

Midstory

Low Severity <30% Consumed Likely patches of unburned fuels, scorched, and charred foliage, with much of the plant structures sill identifiable	**Moderate Severity** 30-75% Consumed Most of the foliage consumed on individual plants, with a range of shrub and small tree skeleton	**High Severity** >75% Consumed All foliage consumed and many shrub and tree skeletons

Midstory Observations

The mid-story often responds more directly to fire than the overstory and includes many of the ladder fuels that influence vertical fire movement. Using the illustrations and comparison table, estimate the overall fire effects within the **Observation Point (OP)** by looking for recurring patterns across small trees and tall shrubs rather than individual plants.

Estimate the average percent within your OP that is:

- Scorched
- Charred
- Consumed

Understory

Low Severity	**Moderate Severity**	**High Severity**
<30% Foliage Scorched	30-75% Foliage Scorched	>75% Foliage Scorched
Likely patches of unburned fuels, scorched, and charred foliage, with much of the plant structures sill identifiable	Most of the foliage consumed on individual plants, with a range of shrub and small tree skeleton	All foliage consumed and many shrub and tree skeletons

Understory Observations

The understory often shows the most visible changes during prescribed and cultural burns. Using the illustrations and comparison table, estimate the overall fire effects within your **Observation Point (OP)** by observing live saplings, shrubs, grasses, and other low vegetation. This layer often reflects how effectively fire met objectives for reducing ladder fuels and modifying understory vegetation.

Estimate the average percent within your OP that is:

- Scorched
- Charred
- Consumed

Record any additional observations that help describe the overall condition of the understory, particularly where fire effects vary across different fuel models (vegetation types).

Surface/Substrate

Low Severity	Moderate Severity	High Severity
<50% Consumed	50-80% Consumed	>80% Consumed
Litter partially blackened, duff nearly unchanged, wood/leaf structures unchanged	Litter charred to partially consumed, upper duff layer burned, wood/leaf structures charred, but recognizable	Litter and duff consumed, leaving fine white ash. Mineral soil visibly altered, often reddish in color

Surface / Substrate Severity

Surface and substrate observations describe how fire affected the fuel bed and soil surface. Using the illustrations and comparison table, estimate the overall fire effects within your **Observation Point (OP)** by observing litter, grasses, needles, woody debris, logs, duff, and exposed soil. This layer typically includes fuels ranging from 1-hour through 1,000-hour time-lag classes.

Observe within your OP:

- Estimated percent of the surface affected or consumed
- Amount of char and remaining plant material
- Presence of white ash, which may indicate more complete combustion
- Remaining duff depth and the amount of duff consumed

If pre-burn observations were not made, compare burned areas with nearby locations that experienced minimal fire effects or retained similar fuel conditions. Where ash or duff has been disturbed, note the color and condition of the exposed soil as an additional indicator of fire effects.

Soil Hydrophobicity

Soil hydrophobicity is often a delayed fire effect and may not become apparent until after rainfall. When conditions allow, simple field observations—such as placing a few drops or pouring a small amount of water onto the soil surface—can provide early indications of water repellency. Record whether water readily infiltrates the soil or beads and runs across the surface, and compare observations among different locations within your **Observation Point (OP)**.

When hydrophobic conditions are present, note their apparent extent and relationship to burn severity, vegetation, slope position, and remaining ground cover. Follow-up observations after rainfall often provide the most meaningful assessment of hydrophobic soil conditions.

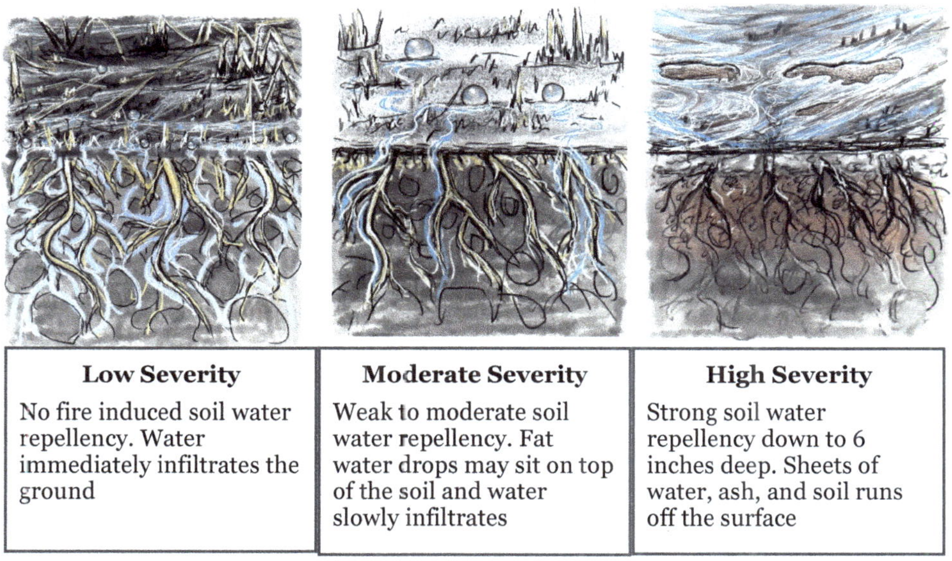

Low Severity	**Moderate Severity**	**High Severity**
No fire induced soil water repellency. Water immediately infiltrates the ground	Weak to moderate soil water repellency. Fat water drops may sit on top of the soil and water slowly infiltrates	Strong soil water repellency down to 6 inches deep. Sheets of water, ash, and soil runs off the surface

Fire Effects Summary

Fire effects are often summarized using qualitative categories such as **No Effect**, **Low**, **Moderate**, and **High**. These categories provide a convenient way to compare observations across the vertical arrangement of vegetation, but they are only summaries of the measurements you have already made. The observations themselves—such as canopy scorch, vegetation consumption, char height, and remaining duff depth—provide the detail needed to understand how fire affected the landscape.

The summary table adapts field observations from the **U.S. Forest Service Field Guide for Mapping Post-Fire Soil Burn Severity (2010)**, the **BAER Soil Burn Severity Field Guide**, and the **National Park Service Thematic Burn Severity Mosaic (1991)** into a simplified framework for field journaling.

Use the summary table as a quick reference for organizing observations across the overstory, mid-story, understory, and surface layers before comparing them with burn objectives. The goal is not to assign a single severity rating to the burn, but to build a consistent picture of how different parts of the vegetation and fuel profile responded to fire.

Comparing Effects with Burn Objectives

The most meaningful evaluation of fire effects comes from comparing your observations with the intended outcomes of the burn. Similar fire effects may represent excellent results for one prescription and undesirable results for another. Burn objectives provide the context needed to interpret your observations.

Overstory	Mid-Story	Understory	Surface
Low Intensity (Scorched) —No to little overstay canopy scorch and high tree survival	**Low Intensity (Scorched)** —0-25% midstory trees and tall shrub canopy scorched	**Low Intensity (Consumed)** —0-25% sapling trees and small shrubs consumed	**Low Intensity (Consumed)** —10-40% litter —30-60% fine (1 hour) fuels —0-10% duff
Moderate Intensity —Some overstory tree scorch, moderate stress to trees	**Moderate Intensity** —25-60% midsotry trees and tall shrub canopy scorched	**Moderate Intensity** —25-60% sapling trees and small shrubs consumed	**Moderate Intensity** —40-70% litter —60-85% fine (1 hour) fuels —10-40% duff
High Intensity —High scorch, potential overstory tree mortality. —Different tree species can survive with different levels of scorch	**High Intensity** —60-90% midstory trees and tall shrub canopy scorched	**High Intensity** —60-90% sapling trees and small shrubs consumed	**High Intensity** —70-95% litter —85-100% fine (1 hour) fuels —40-70% duff —Mineral soil exposed

Some objectives emphasize greater vegetation consumption or tree encroachment reduction, while others seek to retain more vegetation, protect overstory trees, or minimize soil disturbance. Rather than focusing only on severity categories, compare the actual observations—such as percent consumption, canopy scorch, or char height—with the desired range established in the burn plan or other prescribed-fire guidance.

Fuel Models	Percent Consumed/Burn Objectives			Observed
GR	60%	61-89%	90%	30-60%
TL	30%	61-89%	90%	10-30%

The comparison table illustrates one way to relate observed fire effects to burn objectives for different fuel models (vegetation types). This approach keeps the emphasis on observations first and severity categories second, helping your journal become a practical tool for evaluating prescribed fire outcomes.

Journaling Fire Effects

Fire effects invite us to slow down and observe the landscape differently. Unlike fire behavior or smoke, many fire effects remain visible long after the flames have passed, allowing us to compare observations across hours, seasons, and even years. Whether your journal emphasizes measurements, stories, or sensory experiences, each return visit adds another layer to the evolving story of the burn.

The examples that follow demonstrate several ways to journal fire effects. Some emphasize systematic comparisons that support the **Study** approach, while others capture broader ecological patterns and personal experiences through **Story**journaling. Together they illustrate that there is no single way to document fire effects—only different ways of observing change.

Example 1–3: Fire Effects Hourly Trackers

Although fire effects often change more slowly than fire behavior, hourly observations made during and immediately after the burn can help connect fire behavior with its resulting effects. The three examples on the next page illustrate different ways of organizing fire-effects observations within an hourly tracker.

Use consistent time intervals to compare fire effects with weather, fire behavior, and smoke observations. Simple vegetation sketches, symbols, and color coding are often more effective than detailed drawings for visualizing changing patterns.

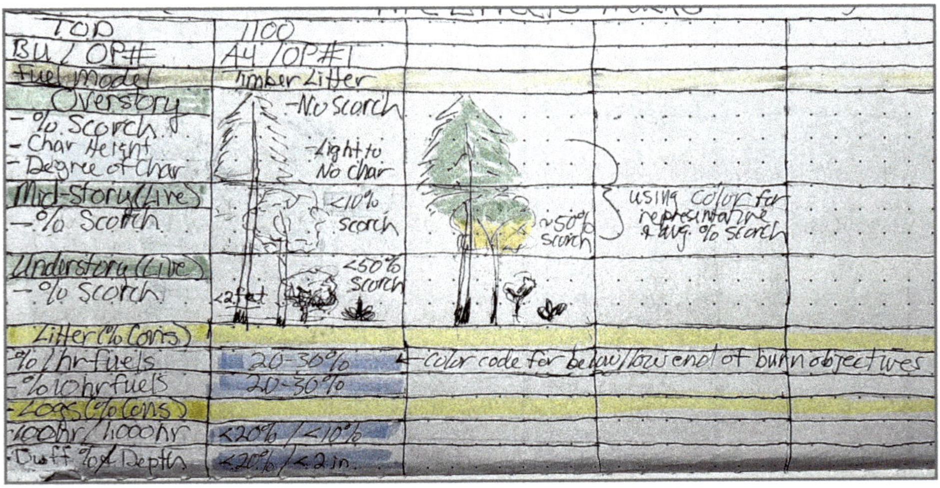

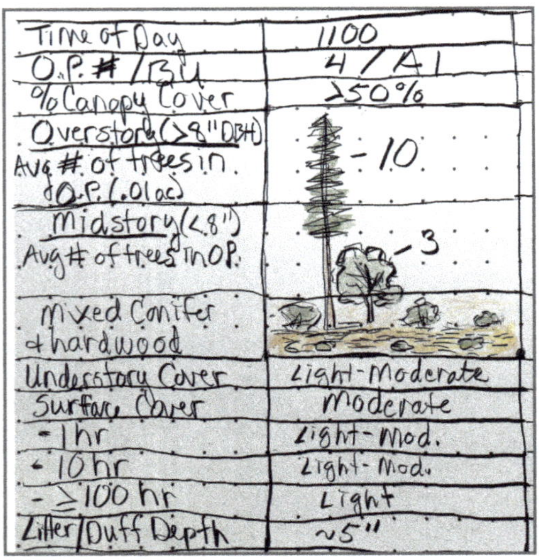

The first column in this hourly tracker was filled in as an example.

The two examples were early versions of how I journaled fire effects for an hourly tracker. The top example is specific to fire effects observations while the second is for fuels observations, but could overlay fire effects observations.

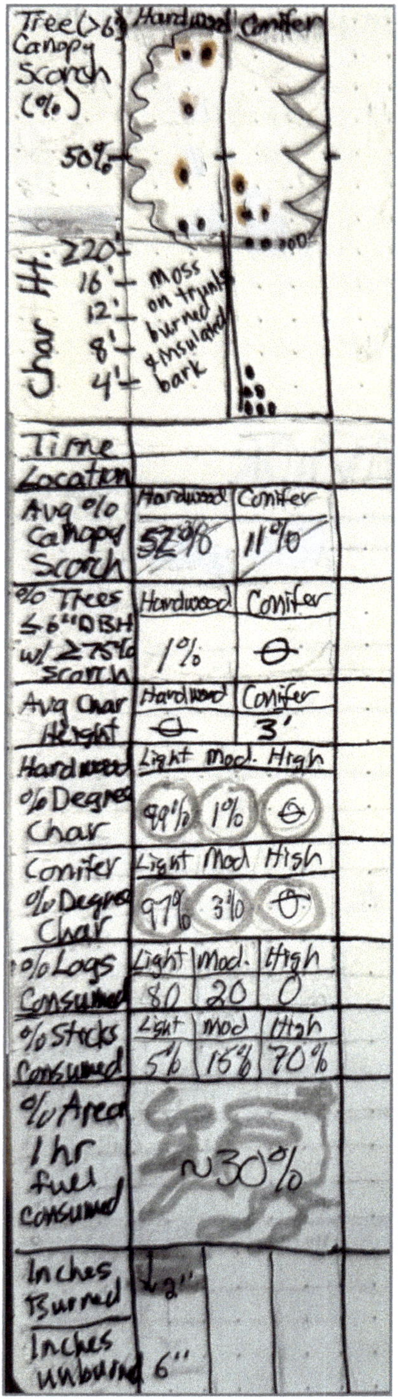

The example to the left shows one hourly column from a newer fire effects tracker that I am developing. The upper portion of the column uses a symbolic tree, with the trunk dividing a hardwood canopy (left) from a conifer canopy (right). Measurement increments along the side of the tree allow canopy scorch to be estimated, while dots within each canopy represent the number of trees and their approximate scorch heights observed within the **Observation Point (OP)**.

As you move down the hourly column, measurements are combined with simple visual symbols to make key observations easier to recognize at a glance. For example, the circles labeled **Low**, **Moderate**, and **High** represent tree trunk diameters, while the space within each circle is used to record the average degree and depth of bark and cambium char observed within the OP. Near the bottom of the column, a horizontal surface-fuel graphic provides a quick visual estimate of the percent and pattern of 1-hour fuel consumption across the observation point.

Thoughtful data visualization makes complex observations easier to interpret and helps reveal patterns and relationships that might otherwise be overlooked.

Example 4–6: Story Journaling Fire Effects

Not every journal page needs to organize observations into measurements or tables. Sometimes a single sketch, comparison, or written reflection captures the overall character of a burn more effectively than detailed data.

The following examples emphasize broader observations of fire effects, comparing vegetation, fuels, and ecological changes rather than documenting hourly patterns. These pages combine sketches, written notes, measurements, and sensory observations to tell the story of how the landscape changed.

In the example above, the fire effects sketch on the left page was very intentional. I chose a scene that retained unburned vegetation in the foreground, with shrub skeletons and ash in the mid-ground to emphasize the contrast created by the fire. Rather than focusing on char height or canopy scorch, my goal was to convey the overall pattern and character of surface fuel consumption.

Simple pencil or pen sketches with a little color work very well for visually accentuating fire effects.

Fire Effects Maps & Overlays

Fire effects can also be viewed spatially across the burn unit. Transparent overlays placed on a burn unit map allow fire effects to be compared with ignition patterns, fire behavior, terrain, and fuel models (vegetation types).

Rather than mapping every observation, use simple dots, patches, or shaded areas to represent general fire effects or severity across the landscape. You may choose to develop a separate fire-effects overlay or build directly from your fire behavior map by adding notes that describe the resulting vegetation consumption, scorch, or other observed changes. Consistent colors and symbols help reveal patterns that might otherwise be overlooked.

A Walk In The Ash

Fire effects are experienced as much through the senses as through measurements. As the landscape quiets, subtle sounds, textures, scents, and visual contrasts often become easier to notice. These observations complement the study approach by adding details that deepen our understanding of how the burn transformed the landscape.

Continue using the deep listening exercise introduced in the Fuels chapter, but compare the sounds of burned and unburned ground. Notice differences as you walk across bare soil, ash, charred vegetation, remaining litter, and unburned fuels.

Listen for observations such as:

- Cracking branches or falling snags
- Hollow footfalls on deep ash
- Changes in footing between burned and unburned vegetation
- Subtle metallic sounds from cooling bark (Michaletz & Johnson, 2012; Baskin & Baskin, 2014)

These sensory observations contribute to both situational awareness and long-term ecological understanding.

Footstep Sound Exercise

Walk slowly through an area with varied fire effects, listening to each footfall. Sketch simple footprint shapes in your journal, arranging them to represent the path you walked through the landscape. As you move, record notes beside individual footprints or develop a simple legend that relates colors to the sounds and conditions beneath each step.

Use colors or symbols to distinguish among unburned vegetation, charred fuels, white ash, bare soil, litter, or other surface conditions. Notice how the sounds change as you move between these different fire effects, and record words or phrases that describe each sound.

When your footsteps have been mapped and the sounds documented, add a small sketch of the surrounding scene. Then review the sequence of footsteps and the rhythm they created across the page. Use musical notes, sound lines, descriptive words, or other visual sound symbols to create a visual soundtrack of the fire effects landscape. Rather than illustrating only what the burn looked like, this exercise captures how it felt and sounded to walk through the changing mosaic of fire effects.

crackle
snap
Crackle
soft thud
soft thud
scratchy sharp crack
scratchy sharp crack
crackle
crackle snap
crunch
crackle
crackle snap
crackle
crackle
crackle
crackle snap
crackle snap
Gray-white ash
Charred pine litter
Partially charred
scorched litter
unscorched litter

silence & light smoke
Pause
snap
crackle crackle
soft thud, soft thud.
PP crackle crackle ff crackle ff
sharp scratchy crackle
PP crackle crackle
snap
sharp scratchy crackle
ff snap crackle
snap crackle snap
Pause

Interactive Journal Pages

As your journal grows, it can become more than a collection of individual pages. Folding flaps, removable inserts, transparent overlays, and layered sketches allow related observations to be viewed together without constantly turning pages. These interactive layouts make it easier to compare weather, fire behavior, smoke, and fire effects while revealing patterns and relationships that may not be obvious when observations remain separated.

The examples that follow illustrate two approaches I frequently use. The first uses removable or flip-up pages to organize forecasts, burn parameters, weather observations, and hourly trackers so they can be viewed alongside sketches and journal entries. I often prepare these pages before the burn using matching sizes and layouts, then attach them with double-sided tape so they fold outward without covering other observations.

The example above uses transparent overlays placed over maps or landscape sketches to layer observations such as ignition patterns, fire behavior, smoke movement, or fire effects. Because each layer remains separate, you can compare changing conditions through time while still seeing the underlying landscape. Together these techniques transform your journal from a sequence of pages into a dynamic visual workspace for studying patterns, relationships, and the unfolding story of the burn.

Returning

One of the greatest gifts of prescribed and cultural fire is that the story does not end when the burn is complete. Fire effects continue unfolding long after the last flame has disappeared. Some changes are visible immediately, while others emerge only after rainfall, new growth, flowering, insect emergence, migration, or the passing of another season.

Returning to the same **Observation Point (OP)** allows your journal to become more than a record of a single day. Each visit reveals something the previous one could not. A hillside that seemed quiet after the burn may later fill with wildflowers, insects, bird songs, fungi, or new seedlings. The ash that once softened your footsteps becomes incorporated into the soil. Charred shrubs begin to resprout, while snags provide new nesting and foraging opportunities. Every return adds another chapter to the story of the burn.

There is no perfect schedule for revisiting a burned landscape. Return when you can, and whenever the seasons invite new observations. Even brief visits reveal relationships that cannot be understood from a single burn day, reminding us that fire effects are not a moment in time but an ongoing process of ecological change.

Growing Your Observation Library

Every journal page becomes part of a growing observation library. Sketches, measurements, maps, questions, and reflections are more than memories—they become references that sharpen future observations. As your library grows, you begin recognizing recurring patterns across different fuel models (vegetation types), seasons, weather conditions, and burn objectives.

You may find yourself recalling the sound of footsteps through deep ash, the shape of a lofting smoke plume, the rhythm of bird songs before ignition, or the way morning light revealed subtle canopy scorch. Experiences that once seemed unrelated begin to connect, allowing each new burn to build upon those that came before.

Over time, your journals become more than records of prescribed fires. They become a visual history of your relationship with fire, helping you recognize change, deepen your understanding of ecological relationships, and continue learning from every landscape you visit.

The Continuing Dance

The Earth is always moving. Water circulates. Carbon is captured, stored, released, and recaptured. Plants grow, decay, and regenerate. Animals move across the landscape while weather and seasons continually reshape the conditions of life. Whether we notice it or not, the ecosystem is always changing—always dancing.

Within that larger ecological dance, vegetation forms a living dance troupe. Trees, shrubs, grasses, litter, and duff each play their part, growing, competing, dying, and returning as the landscape continually renews itself.

Fire does not begin the dance; it joins it. When fire becomes a dance partner, transformation unfolds. Stored energy is released, heat is exchanged, and relationships among weather, terrain, fuels, wildlife, and people are rearranged. Fire behavior is one powerful movement within the continuing choreography of the ecosystem.

As the fuels dance with fire, some are transformed into smoke—a great exhalation of the ecosystem that rises, drifts, settles, and eventually returns to the earth. Other fuels remain, forever changed by the encounter. Together they reveal both the history that accumulated before the burn and the beginnings of a new story.

When the flames have passed, the dance does not end. The dancers have changed. Some remain standing. Some return from roots or seed. New dancers arrive. Birds establish territories. Insects emerge. Fungi fruit. Flowers bloom. The choreography continues as relationships are renewed, reshaped, and rediscovered with every passing season.

As journalers, we become participants in that continuing dance. We do more than record observations—we learn to notice relationships, recognize patterns, and witness transformation through time. Every sketch, measurement, question, and story becomes another step toward understanding the living rhythms of fire and the landscapes it helps shape.

The landscape is already dancing.

Bring a spark of curiosity—and your journal.

Acknowledgments

This book grew from countless hours spent on prescribed burns, cultural fire discussions, nature journaling outings, workshops, and conversations with people who have generously shared their knowledge, experience, and curiosity. I am deeply grateful to the many fire practitioners, scientists, educators, artists, land stewards, and journalers whose ideas have shaped both my understanding of fire and the pages of this book.

I owe a special thanks to the prescribed fire and cultural fire practitioners who continue to demonstrate that good fire is built not only upon sound science and careful planning, but also upon relationships—with one another, with the landscape, and with the living communities that fire helps sustain. Your willingness to teach, question, and share your perspectives has continually broadened my own.

I am equally grateful to the many ecologists, researchers, and authors whose work has deepened my appreciation for the intricate relationships among weather, terrain, fuels, wildlife, and people. Their research provides the scientific foundation upon which my own observations continue to grow.

Finally, thank you to every journaler who has ever paused to look a little longer, ask another question, sketch one more page, or return to a landscape after the fire was over. Curiosity is contagious. Every observation adds another thread to our collective understanding of good fire.

A Special Thanks To....

Delphine Griffith, for inviting me to teach and participate in several fire effects monitoring training events, which became the initial spark that inspired this book.

Jessie Thoreson, for partnering with me on numerous fire illustration and journaling projects, bringing fresh creativity, enthusiasm, and energy to Pyrosketchology.

My husband, Kipp Morrill, whose unwavering encouragement, patience, and support have made every Pyrosketchology adventure—including this book—possible.

References

Many concepts throughout this book are adapted from fire management guidance, ecological literature, and nature journaling practices. These references provide additional resources for readers interested in exploring these subjects in greater depth.

Books

Anderson, M. K. (2005). Tending the Wild: Native American Knowledge and the Management of California's Natural Resources. University of California Press.

Baskin, C. C., & Baskin, J. M. (2014). Seeds: Ecology, Biogeography, and Evolution of Dormancy and Germination (2nd ed.). Academic Press.

Haskell, D. G. (2017). *The Songs of Trees.* Viking.

Kimmerer, R. W. (2013). Braiding Sweetgrass: Indigenous Wisdom, Scientific Knowledge, and the Teachings of Plants.Milkweed Editions.

Leopold, A. (1949). *A Sand County Almanac.* Oxford University Press.

Martin, R. E., & Hamman, S. T. (2016). *Prescribed Fire and Fuel Management.* University of California Agriculture and Natural Resources.

Morrill, M. H. (2024). Pyrosketchology: An Illustrated Guide to Nature Journaling About the Fire Environment. Self-published.

Morrill, M. H. (2024). "Pyrosketchology." In *Elementals: Fire* (Vol. 4). Center for Humans and Nature.

Odum, E. P., & Barrett, G. W. (2005). *Fundamentals of Ecology* (5th ed.). Brooks/Cole.

Tallamy, D. W. (2020). Nature's Best Hope: A New Approach to Conservation That Starts in Your Yard. Timber Press.

Fire Science & Ecology

Andrews, P. L., Bevins, C. D., & Seli, R. C. (2011). BehavePlus Fire Modeling System, Version 5.0: User's Guide. USDA Forest Service, Rocky Mountain Research Station.

Campbell, D. (2004). A Strategy for Fire Behavior Prediction. Wildfire Magazine.

Cruz, M. G., Alexander, M. E., & Wakimoto, R. H. (2019). "Modeling the Rate of Spread of Wildfires." *International Journal of Wildland Fire*.

Cruz, M. G., Alexander, M. E., & Wakimoto, R. H. (2022). "Advances in Fire Spread Modeling." *International Journal of Wildland Fire*.

Hewson, E. W., Wade, J. E., & Baker, J. (1979). The Use of Trees as Indicators of Wind Power Potential. Oregon State University / U.S. Department of Energy.

Long, L. L., Lake, F. K., Stephens, J. L., Alexander, J. D., Ralph, C. J., & Wolfe, J. D. (2023). Using Culturally Significant Birds to Guide the Timing of Prescribed Fires in the Klamath-Siskiyou Bioregion. Ecosphere, 14(6), e4541. https://doi.org/10.1002/ecs2.4541

Michaletz, S. T., & Johnson, E. A. (2012). "How Forest Fires Kill Trees: A Review of the Fundamental Biophysical Processes." *Scandinavian Journal of Forest Research*.

Rothermel, R. C. (1972). *A Mathematical Model for Predicting Fire Spread in Wildland Fuels* (Research Paper INT-115). USDA Forest Service, Intermountain Forest and Range Experiment Station.

Scott, J. H., & Burgan, R. E. (2005). Standard Fire Behavior Fuel Models: A Comprehensive Set for Use with Rothermel's Surface Fire Spread Model (RMRS-GTR-153). USDA Forest Service, Rocky Mountain Research Station.

Fire Management References

National Interagency Fire Center & National Wildfire Coordinating Group. *NWCG Glossary of Wildland Fire Terminology*.

National Wildfire Coordinating Group. *Fire Effects Monitoring Handbook*.

National Wildfire Coordinating Group. *Fireline Handbook*.

National Wildfire Coordinating Group. *Incident Response Pocket Guide (IRPG)*.

National Wildfire Coordinating Group. Prescribed Fire Complexity Rating System Guide.

National Wildfire Coordinating Group. Prescribed Fire Planning and Implementation Procedures Reference Guide (PMS 484).

National Wildfire Coordinating Group. *S-290 Intermediate Fire Behavior*.

National Wildfire Coordinating Group. Smoke Management Guide for Prescribed and Wildland Fire (PMS 420-2).

National Wildfire Coordinating Group. *Wildland Fire Modules Guide*.

U.S. Forest Service. (2010). Field Guide for Mapping Post-Fire Soil Burn Severity.

U.S. Forest Service, Rocky Mountain Research Station. *Wildland Fire in Ecosystems* (General Technical Report RMRS-GTR-42 series).

U.S. Forest Service, Southern Research Station. *Prescribed Fire in Southern Forests*.

U.S. National Park Service. Thematic Burn Severity Mosaic of the Conterminous United States (CONUS).

Additional Resources

Natural Resources Network. Common Prescribed Burning Parameters in Northeastern Fuels.

Oregon Department of Environmental Quality. (2023). *Air Quality Index and Visibility Guidance*.

USDA Natural Resources Conservation Service. *Soil Biology Resources*.

Xerces Society. Pollinator Conservation Resources.

Yedinak, K. M. Fire Acoustics Research and Training Materials.

About the Author

Miriam H. Morrill is a retired wildland fire professional, author, illustrator, and educator dedicated to advancing fire communication through nature journaling, visual storytelling, and observation. She is the author and illustrator of *Pyrosketchology: An Illustrated Guide to Nature Journaling About the Fire Environment* and the essay **"Pyrosketchology"** in *Elementals: Fire* (Volume 4 of the *Center for Humans and Nature* anthology series).

During her 27-year federal career with the U.S. Forest Service, Bureau of Land Management, and U.S. Fish and Wildlife Service, Miriam served in a wide variety of field, state, regional, and national roles, including firefighter, wildlife biologist, soil, water, and air specialist, climate change adaptation planner, public affairs specialist, and ultimately California State Fire Prevention, Mitigation, and Education Specialist.

Her experience includes participation on numerous national and international fire response and training teams. She has supported fire prevention and investigation training in Jamaica, community wildfire protection planning in Micronesia, the Republic of Palau, and Guam, and has provided scientific illustration and visual communication support for fire scientists and educators in the United States, Canada, and the Netherlands, including contributions to the 2023 Global Fire Science Conference.

Miriam has designed and led Pyrosketchology workshops for The Nature Conservancy's Prescribed Fire Training Exchange (TREX) Program, the National Fire Learning Network, Fire Adapted Communities, Fire Safe Councils, the International Association of Wildland Fire, universities, and numerous community organizations. Her workshops emphasize careful observation, curiosity, and visual thinking as practical tools for understanding fire and strengthening relationships with people, place, and the living processes of fire.

For more information about Pyrosketchology, workshops, and resources, visit **pyrosketchology.com**.

www.ingramcontent.com/pod-product-compliance
Lightning Source LLC
LaVergne TN
LVHW050911210926
848482LV00012B/407
* 9 7 9 8 9 9 0 1 0 7 1 2 0 *